CW00501735

INSTITUTION OF CIVIL ENGINEERS

The Millennium Link
The rehabilitation of the Forth & Clyde and Union Canals

Proceedings of the international conference organized by the Institution of Civil Engineers and held in Edinburgh, UK, on 30 June – 1 July 2000

Edited by G. Fleming

Thomas Telford

This conference was organized by the Conference Office of the Institution of Civil Engineers

Organizing Committee:
G. Fleming, *Institution of Civil Engineers*, Chairman
G. Ballinger, *British Waterways*
I. Bruce, *Scottish Enterprise*
O. Lassiere, *British Waterways*
G. Masterton, *Babtie Group*
R. Paxton, *Panel for Historical Engineering Works*
G. Pender, *University of Glasgow*
N. Raufaste, *American Society of Civil Engineers*
G. Shilebeer, *Enviro Centre*
J. Stirling, *British Waterways*
F. Smith, *University of Dundee*
J. Walker, *Scottish Enterprise*

Conference sponsors:
American Society of Civil Engineers
Association of Municipal Engineers
British Waterways
Institution of Civil Engineers
Scottish Enterprise

Cover photograph courtesy of British Waterways.

Published by Thomas Telford Publishing, Thomas Telford Ltd, 1 Heron Quay, London E14 4JD.
URL: http://www.t-telford.com

First published 2000

Distributors for Thomas Telford books are
USA: ASCE Press, 1801 Alexander Bell Drive, Reston, VA 20191-4400, USA
Japan: Maruzen Co. Ltd, Book Department, 3–10 Nihonbashi 2-chome, Chuo-ku, Tokyo 103
Australia: DA Books and Journals, 648 Whitehorse Road, Mitcham 3132, Victoria

A catalogue record for this book is available from the British Library

Availability: Unrestricted
Content: Collected papers
Status: Refereed
User: Civil engineers, local authorities

ISBN: 0 7277 2945 4

© The Institution of Civil Engineers, unless otherwise stated, 2000

All rights, including translation, reserved. Except for fair copying, no part of this publication may be reproduced, stored in a retrieval system or transmitted in any form or by any means, electronic, mechanical, photocopying or otherwise, without the prior written permission of the Books Publisher, Thomas Telford Publishing, Thomas Telford Ltd, 1 Heron Quay, London E14 4JD.

This book is published on the understanding that the authors are solely responsible for the statements made and opinions expressed in it and that its publication does not necessarily imply that such statements and/or opinions are or reflect the views or opinions of the publishers or of the Institution of Civil Engineers.

Printed and bound in Great Britain by Bookcraft (Bath)

Contents

The canals of Scotland: an historical overview

JOHN R. HUME

Scotland's canals can be divided into three groups: ship canals, designed in whole or in part for through passage by masted vessels; barge canals designed for public use, and barge canals intended for use by a single user. These waterways, as elsewhere, complemented existing river navigations, and the ship canals, as might be expected, were aids to longer-distance trade and communication. The 'system' thus created was only in a very limited sense a network, though there was some through traffic on the four interlinked lowland canals. The network of English narrow canals, and the river-navigation/barge canal systems had no parallels in Scotland, and there was no tradition in Scotland of families living on boats, and hence no tradition of the canal 'culture' to be found on the English narrow system. This was largely a function of the relatively short distances involved. The main internal traffics on the lowland Scottish canals never involved a one-way journey of more than a few hours, or more than a single overnight stay for a return trip. These differences from England are important to an understanding of the peculiarities of the Scottish inland waterways.

The most significant of the three groups are the ship canals. There are three of these: in date order the Forth and Clyde, the Crinan and the Caledonian. The geography of Scotland is such that the land mass of the country is almost divided into three by the estuaries of the Forth and Clyde, which penetrate deep into the central lowlands, and further north by Loch Linnhe and the Moray Firth, at either end of the rift valley known as the Great Glen, which is itself punctuated by three fresh water lochs. The relatively narrow land barriers at both of these locations made them natural foci for ship canal schemes. So too, on a much smaller scale, did the neck of land between Loch Gilp and Loch Crinan, at the head of the Kintyre peninsula on the west coast of Scotland. As until the early 19th century most of the trade of Scotland was carried on in relatively small sailing vessels, the notion of a ship canal was not too intimidating, and the advantages of building such canals clear. Not only would a canal at any of these locations reduce the distance to be traversed very significantly, but it would much reduce the risk from both natural and man-made hazards. The voyage round the north of Scotland involved passage through the Pentland Firth, where the Atlantic and North Sea tides meet dangerously. The journey round the Mull of Kintyre also exposed craft to conflicting tides and to the full force of Atlantic swells. In time of war an additional hazard was the risk of privateers on the west coast, and the risk of attack by enemy warships on the east.

Of the three routes mentioned, by far the most commercially attractive was that between the Firth of Forth and the Firth of Clyde, as it reduced voyage distances much more than a canal through the Great Glen, and also linked the two important centres of population at either end of the central lowland valley, centred on Glasgow and Edinburgh, but with extensive overseas and coastal trading links. The first serious proposal for a Forth-Clyde canal seems to have

The Millennium Link. Thomas Telford, London, 2000, 1–6

been made in the reign of Charles II, and may well have been stimulated by the construction of the major French Canal du Midi, as part of Louis XIV's improvements in transportation. Certainly a survey was made by James Smith, better known as an architect, for such a canal, but no construction was undertaken. The idea was revived in the 1740s, perhaps influenced by the building of the Newry Canal in northern Ireland, to link Lough Neagh with Carlingford Lough, but again without effect. The construction and commercial success of the Duke of Bridgewater's canal from Worsley to Manchester, opened in 1761, was probably the trigger for revived interest in a Forth-Clyde link, but by that time the industrialisation of central Scotland was also underway, and the large integrated ironworks at Carron was anxious to gain easy access to the west of Scotland, both as a market in its own right, and as a means of entry to the West Indies and southern American colonies, sharing these interests with the Leith merchants. At the other end, the Glasgow tobacco merchants wished to get safer and speedier links to their main markets on the Continent.

The east and west-coast interests had different perspectives on what kind of canal they wanted. The Glasgow men wished to have a shallow barge canal to take their tobacco to the east coast for transfer to sea-going vessels. This would deny the east-coast traders direct access to the west by sea-going vessels, which was what they wanted. Eventually the latter view prevailed, and the Act of Parliament of 1768 authorisedthe construction of a ship canal - the Great Canal - as surveyed by John Smeaton. This was intended to run from the mouth of the Grange Burn on the Forth to Dalmuir on the Clyde, with a branch to Glasgow. Construction began at the east end, where the town of Grangemouth grew up, and reached Kirkintilloch in 1773, having crossed the difficult Dullatur Bog. At Kirkintilloch a large aqueduct was built over the river Luggie, and then the canal was cut to the outskirts of Glasgow, at Stockingfield, which it reached in 1775. There the capital ran out. The Glasgow merchants raised enough money to complete a branch to a basin at Hamiltonhill in 1777. and there work stopped until 1784, when the Government agreed to make money available from the Forfeited Estates Fund as a loan to allow the canal to be completed. By this time the resident engineer for the eastern section, Robert Mackell, was dead, and the new works were entrusted to Robert Whitworth. He substantially revised the original proposal, taking the western terminus of the canal further down the Clyde to Bowling, and using the Monkland Canal (see later) to convey water from new reservoirs at its eastern end to the new Glasgow terminus of the Forth and Clyde at Port Dundas. This involved completing the Monkland, with locks at Blackhill, and making a 'Cut of Junction' from its Glasgow end to the Port Dundas basin. The largest works on the link to Bowling were a flight of five locks, with oval basins in between, a drydock, and an aqueduct over the river Kelvin, all at Maryhill, near Glasgow. The aqueduct was said to be the largest in Europe at the time. The completed canal was opened in 1790, and appears to have been the first sea-to-sea ship canal of modern times.

The Forth and Clyde was a great success. It built its own railway to connect its docks at Grangemouth with the main-line system. It was the lure of a coal-exporting port on the Forth that led the Caledonian Railway to buy the Forth and Clyde Canal Company in 1867. Under railway ownership the canal continued an important trade in timber to Glasgow from Grangemouth. Other significant canal-borne trades were coal from canal-side collieries, whisky, and grain. From the 1890s to 1939 canal trips from Glasgow to Craigmarloch, near Kilsyth, were also popular. Latterly fishing boats and yachts used it as a through route. By the early 1960s it was little used, and the construction of the A80 Glasgow-Stirling trunk road was used as the reason for abandoning navigation in 1962. Thereafter the canal was partly infilled and blocked by road crossings. Thesection through Clydebank was also shallowed.

The canal was however protected from further fragmentation by a Subject Plan approved by all the local authorities through which it passed, and restoration began in the late 1980s. Planning undertaken at that time was an essential background to the success of the Millenium Link project.

The second ship canal was the Crinan, designed by John Rennie, and built under an Act of 1794. It is a much more modest affair than the Forth and Clyde, but even so was only completed in 1801 with Government aid. It was built with locks slightly longer than those of the Forth and Clyde, so that it extended the range of the vessels which could use the latter. It runs from Ardrishaig on Loch Gilp to Crinan on Loch Crinan.

The last of the ship canals was the Caledonian. This was proposed by Thomas Telford in a report commissioned by Government on the transport needs of the Highlands, presented in 1803. His recommendation was for the construction of a waterway capable of taking small sailing warships as well as trading vessels. It was vastly larger in its width and depth, and in the length of its locks, than either of its predecessors, and indeed was the largest canal then designed. Because of the existence of the lochs in the Great Glen it consisted of four sections of artificial waterway linking these lochs to Loch Linnhe in the west and the Beauly and Moray firths in the east, though the central loch, Oich, was so shallow that a navigable channel had to be dredged through it. The construction of the sea-locks at either end proved difficult, as did the Fort Augustus flight of locks at the south west end of Loch Ness, which required three large steam pumping-engines to keep the water at bay in the bottom lock chamber. Telford and his resident engineers used railways and cranes to expedite the construction, which was completed in 1822, though subsequently some of the structures had to be upgraded.

Both the Crinan and the Caledonian were adversely affected by the advent of the steamship, especially at first the Crinan, as steamers quickly outgrew the size of its locks. Apart from small steam coasters - 'puffers' - designed for the most part so that they could use the Forth and Clyde, the Crinan became a link between steamers serving its terminal basins, and cargo vessels avoided the canal altogether as the Mull of Kintyre lost its danger and the risk of attack from privateers vanished. The Caledonian became a link in a route from the west Highlands to Inverness, served by a succession of paddle steamers built to fit the locks on the canal and connecting at the west end withsteamers from Oban, and from 1894 with a branch from the West Highland Railway. As with the Crinan, however, the increasing size and power of steamships meant that they did not need to, or could not use the canal. It was, however much used by fishing boats transferring between fishing grounds, a highly seasonal traffic.

The Caledonian was built entirely, and the Crinan partly. with Government funding, In 1848 the Crinan was taken over by Government, and managed jointly with the Caledonian. When the waterways were nationalised in 1948 these canals were simply transferred to the British Transport Commission,and as they were still in use became in 1968 'commercial waterways'. This status has ensured that they have never been closed to navigation, alone among Scottish canals. Both have been extensively used by yachtsmen, and in the case of the Caledonian, by hire cruisers, for many years.

The second group of canals mentioned in the introduction, barge canals for public use, were all in the central belt of Scotland wiith the exception of the Aberdeenshire Canal (see below). The earliest was the Monkland Canal, surveyed by James Watt, and built under an Act of

1770 to link coal mines in the north-east of Lanarkshire with Glasgow. As with most Scottish canals there was a shortage of capital, and its first western terminus, opened in 1773, was on the outskirts of what is now Glasgow, beside Barlinnie Prison. Between 1782 and 1785 the canal was extended to Blackhill, and a detached length built at a lower level to a basin at Castle Street, close to the centre of Glasgow. The flight of locks linking the two sections of canal was built under an Act of 1790 which also resulted in an easterly extension to draw water from the river Calder, and a link with the Forth and Clyde canal via the 'Cut of Junction' (see above). These improvements were opened in 1793.

Coal was the mainstay of what was a very profitable canal for much of its life, but the creation of an iron-smelting industry in the Coatbridge area from 1828 proved of the greatest importance to the canal. Four branches were built which served ironworks, bringing in raw materials and taking pig-iron to Glasgow, or to Bowling for shipment in larger sea-going vessels. Both horse-drawn and steam barges (known on the lowland canals as scows) were used. The canal became so busy that the Blackhill locks became a serious bottleneck. In 1841 the locks were doubled, and soon after, in 1846, the company merged with the Forth and Clyde Canal. In 1849-50 an inclined plane was built parallel to the Blackhill locks, to reduce water consumption. This was the only canal inclined plane in Scotland. Two railways were built in connection with the Monkland. The first, the Monkland and Kirkintilloch, linked the Monkland Canal with the Forth and Clyde. It opened in 1826. The other was the Drumpeller Railway, linking coal pits to the canal, and opened in the mid 1840s. Though the main traffic on the canal was freight, there was also a passenger service until the late 1840s, when it succumbed to railway competition. The canal's staple traffic was eroded by exhaustion of local mineral deposits, and by the increasingly pervasive use of railways. Navigation ceased c1933, bu the canal remained intact as a feeder to the Forth and Clyde until the 1960s, when infilling began for road improvement, Much of the route is now under the M8 motorway and under local roads in Coatbridge. The surviving fragments are now well looked after by North Lanarkshire Council

The second public barge canal to be completed was the Aberdeenshire Canal, which was built under an Act of 1796, and opened in 1805. The engineer was John Rennie. It ran from Aberdeen to Inverurie, and was primarily built for transporting lime, coal and dung up-country, and granite, corn and other farm produce to Aberdeen for sale or onward shipment. A passenger service was also provided. The canal suffered from competition from the parallel turnpike road, and was sold in 1848 to the Great North of Scotland Railway, which in 1854 opened part of its route from Aberdeen to Keith, built on the bed of the canal. This was the first such conversion in Scotland.

The next barge canal to be constructed was conceived on a much larger scale. This was the Glasgow, Paisley and Ardrossan Canal, designed by John Rennie to link Glasgow with Paisley and Johnstone, both growing textile towns, and all three with the Earl of Eglinton's new harbour and town of Ardrossan. As was so often the case the capital to complete the waterway was not forthcoming, and the canal was only built as far as Johnstone. It was a level canal, and though its eastern terminus was in Glasgow it did not connect either with the Forth and Clyde or with the Clyde navigation. Its traffic was mainly in small goods, notably parcels of cloth, and it developed in the 1820s a considerable passenger trade as a result of introducing light iron fly boats. This did not, however survive competition from the railway to Paisley and Johnstone, which opened in 1840.

Before that, however, the company had opened its own railway as a partial substitute for the completion of its canal to Ardrossan. This line, from Ardrossan to Kilwinning, was opened in 1828, and was successful in conveying both coal and passengers. The railway split from the canal in 1840.The canal survived as an independent concern until 1869, when it was bought by the Glasgow and South Western Railway which closed it in 1881, and rebuilt it as a railway between then and 1885. It is still known as the 'Canal Line', and still runs over the largest engineering work on the canal, the Blackhall Aqueduct.

The last of the large public barge canals to be built, and the last major canal to be constructed in Scotland, was the Edinburgh and Glasgow Union Canal, usually known simply as the Union Canal. The impetus for the construction of such a canal was a desire to reduce the cost of coal in Edinburgh, by linking the capital with the Lanarkshire coalfield. A number of schemes were drawn up in the 1790s, during the Canal Mania in England, for routes from the Broomielaw harbour in Glasgow to the port of Leith. John Rennie examined these schemes in 1797-98, reducing them to two, a southern one through Midcalder and Cleland, and a northern one through Linlithgow, Falkirk and Cumbernauld. Because the southern route passed through an extensive coalfield. he favoured it, but the capital to build it was not forthcoming. It is interesting to speculate on how its water supply would have been secured. The French Wars, with their demands for funds put an end to this project, but the idea of a link to the Lanarkshire coalfield was revived in 1813 as a result of high coal prices in the capital. This time the scheme was devised by the engineer of the Forth and Clyde Canal, Hugh Baird, and reduced the cost very significantly by linking the new canal to the Forth and Clyde at Falkirk and dropping the idea of links to the Broomielaw and Leith. This met with considerable opposition from proponents of the more ambitious scheme, but Telford reported in favour of Baird's scheme in 1815, and after much wrangling the Baird scheme secured its Act in 1818. His original line was modified by cutting back at the Edinburgh end, and by avoiding passage through the Callendar Estate at Falkirk by cutting Scotland's only canal tunnel. To secure a level route it was also necessary to build three tall aqueducts over the rivers Avon and Almond and the Water of Leith. These were designed by Baird with advice from Telford, and had cast-iron troughs to carry the channel. The canal was supplied with water from Cobbinshaw Loch via the river Calder and a feeder channel. It opened in 1822. Apart from the passenger traffic between Edinburgh and Glasgow, with both day and night services, the main trade was coal and building materials into Edinburgh and parcels of 'merchant goods' from Edinburgh to the towns and villages on the route.

The advent of railways had a marked effect on the Union Canal. A junction with the Slamannan Railway, opened in 1840 was formed at a basin at Causewayend. This provided another route for cheap coal to reach Edinburgh, and also gave, briefly, the quickest way to go between the capital and Glasgow, using the Slamannan, Ballochney and Garnkirk and Glasgow railways. By that time, however, the Edinburgh and Glasgow Railway was under construction. When it opened in 1842. it paralleled the Union Canal for much of its length, and took much of its traffic. It took over the canal in 1849. Commercial traffic continued until 1933, but the canal was not finally closed until 1965, as it supplied water to a number of industrial concerns in the Edinburgh area. After closure it was blocked by a number of roads, and piped where it ran through a new housing area at Wester Hailes. Further piping was, however avoided by public pressure, which led directly to the organisation of a canal enthusiast movement in lowland Scotland, and to local authority concern to secure the surviving canals, thus paving the way for the Millenium Link project.

The last group of canals to be mentioned is the barge canals built for use by a single user or small group of private users. Excluding what were in effect river navigations these were, in alphabetical order, the Burnturk Canal (c1800), the Campbeltown Canal (1794), the Kilbagie Canal (c1780), the Muirkirk Canal (c1789), the Stevenston Coal Canals (1770s on). All these were use primarily for conveying minerals, usually coal, though the Kilbagie Canal served a very large distillery. The Muirkirk Canal also supplied water for power to the Muirkirk Ironworks.

As in England there were many abortive proposals for canals, often to open up new areas for development. In retrospect few if any of these could ever have been profitable. The canals that were built, however were a varied and intriguing group. The sea-to-sea ship canals were genuinely innovative, and the Forth and Clyde and Caledonian canals are arguably of world importance. If the Union canal was less novel, it is nevertheless a remarkably unaltered late canal built on a grand scale, with the kind of earthworks and both major and minor structures which anticipated the first main-line railways. Taken collectively the surviving Scottish canals can stand comparison with any of their peers in every way other than extent. It is for this reason that the restoration of navigation on the Forth and Clyde and Union canals should be warmly welcomed on historical as well as on other grounds.

From decline to a new sustainable future
Project overview, finance and fundraising

J. M. STIRLING, BSc, MBA, CEng, MICE, MIStructE
DIRECTOR, SCOTLAND, BRITISH WATERWAYS, GLASGOW, UNITED KINGDOM.

INTRODUCTION

The Millennium Link project was conceived as a mechanism by which the restoration of the 110 kilometres of the Forth and Clyde Canal and the Edinburgh and Glasgow Union Canal would become the focus for a much broader transformation. The project is the culmination of over 25 years of campaigning by canal-side communities who were major losers from economic restructuring. The £78million regeneration of the waterways is intended to act as a stimulus for wider community and environmental renewal **[Fig.1]**.

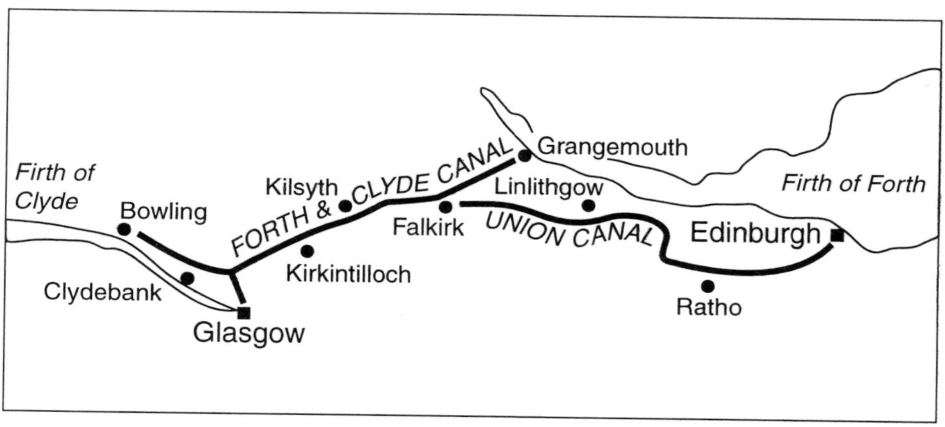

Figure 1. Millennium Link map. [British Waterways]

DECLINE AND CLOSURE

The original construction of the Forth and Clyde Canal as a ship canal allowed the navigation to make an important contribution to the Scottish economy for nearly 150 years. It was the closure of Grangemouth Docks for merchant shipping throughout the First World War, which effectively ended major traffic. The success of the canal led to the creation and growth of many canalside communities. These were mostly connected with the growth of industry beside this new transportation facility, but were also involved in servicing the actual trade along the canal. Kirkintilloch became Scotland's first inland port and developed into a major canal town. The Forth and Clyde provided the City of Glasgow for the first time with a port of its own. The canal became important for passenger traffic with busy trade between

The Millennium Link. Thomas Telford, London, 2000, 7–14

Glasgow and Falkirk from where various stagecoaches transferred passengers to Edinburgh and towns throughout east central Scotland. Once the Edinburgh and Glasgow Union Canal was opened boats conveyed passengers directly to Edinburgh. This passenger traffic was cultivated through the development of faster and more comfortable vessels. In 1836 nearly 200,000 passengers were carried and the journey time between Glasgow and Edinburgh was less than 7 hours[1]. Despite the arrival of railways some passenger traffic continued until about 1883. This was revived in 1893 for pleasure trips from Glasgow and these continued up until 1939, when further wartime restrictions ended that trade.

Traffic on the much smaller Union Canal declined more quickly than on the Forth and Clyde with the company running at a loss from 1842 only 20 years after opening. In 1921 the total tonnage carried was only 20,000. In the same year the terminal basins in Edinburgh were abandoned and sold and in 1933 the connecting flight of locks with the Forth and Clyde at Falkirk were infilled.

As the canals fell into disuse they became targets for both public disquiet about safety and from the local authorities who wished to improve road links by removing restrictive bridges and aqueducts. The canals were seen as having no purpose and their value to the Scottish economy and canalside communities was lost. As a result the Forth and Clyde Canal was closed to navigation on 31st. December 1962 and the Edinburgh and Glasgow Union Canal in August 1965.

Following closure the canals were severed or blocked in over thirty locations. In the simplest cases opening bridges became fixed. In other cases roads were built at grade across new culverts. Lengths totalling 7 kilometres were infilled, the most damaging being over 2 kilometres at Grangemouth and 1.7 kilometres through a large peripheral housing estate at Wester Hailes in Edinburgh. Locks fell into disrepair and a multitude of public utility services were placed across the canals.

VALUE REDISCOVERED

Local pressure contributed to closure, but attitudes changed with new local pressure to improve the amenity and safety of the canals. Some of this pressure came from canal enthusiasts, but importantly a great deal came from canalside communities themselves. This local pressure led to official action and as early as 1976, less than 14 years following closure, Strathclyde Regional Council in its Regional Report identified the rehabilitation of the Forth and Clyde Canal for recreational purposes to be of regional significance. This change in attitude led through a series of stages to the formal adoption in November 1988 of a unique document – The Forth and Clyde Canal Local Plan[2]. This statutory plan is unusual in that it is linear and crosses local authority boundaries. Its policies protected the canal against any further obstructions to navigation and supported restoration as finances allowed.

The will for restoration was growing, but not the means. Some piecemeal improvements were implemented during the late 1980's and early 1990's, when fixed bridges giving navigable headroom replaced four culverted crossings. A long-term project of lock gate renewal was also begun at that time. With the onset of the National Lottery in 1994 it became realistic to consider full restoration from Grangemouth on the Forth to Bowling on the Clyde and between Glasgow and Edinburgh and The Millennium Link project was born.

THE MILLENNIUM LINK – SUSTAINABLE DEVELOPMENT

British Waterways accepted the lead role as project promoters and project managers. The civil engineering to restore both canals is challenging and complex. Apparently simple new bridge crossings involve difficult road geometry, complicated service diversions and important local planning issues. The tight budgetary control, programming and procurement issues are also familiar **[TABLE 1]**. It is important not to diminish the significance of these challenges nor the hard work and dedication required to achieve engineering success. Other papers throughout the conference will highlight the issues and challenges being overcome

CONTRACT	SITE	SCOPE	FORM OF CONTRACT
1 £3m.	A801 Lathallan Rd. Vellore Road Greendykes Rd M8 Broxburn	Fixed bridge Fixed bridge Fixed bridge Fixed bridge and realigned canal	ICE Design & Construct
2 £2m.	Townhead Cadder Balmuildy	Fixed bridge Raise bridge Fixed bridge	NEC/ECC, Option C, Target contract with Activity Schedule, Design and construct
3 £4.5m.	Cleveden Road Cloberhill pipes Duntreath Avenue	Fixed bridge Fixed bridge Fixed bridge	ICE Design & Construct
4a £2.5m.	Wester Hailes-Phase1	Canal and bridges	NEC/ECC, Option A - Priced contract with Activity Schedule, Design and construct
4b £7m.	Wester Hailes-Ph2	Canal and bridges	NEC/ECC , Option A, Design and construct
6 £17.5m.	Falkirk Interchange	Canal, tunnel,aqueduct and wheel	NEC/ECC , Option A, Design and construct
7 £5m.	Lock 16 Lock 11 Camelon Road Merer's bridge Bainsford	Fixed bridge Fixed bridge and new lock Fixed bridge and new lock Fixed bridge and new lock	NEC/ECC , Option A, Design and construct
8 £4.5m.	Grangemouth Link	Canal and 2 locks	NEC/ECC , Option A, Design and construct
9 £5m.	Dalmuir pipes Kilbowie Road Argyll Road	Fixed bridge and droplock Fixed bridge Fixed bridge	NEC/ECC , Option A, Design and construct
10a £1.5m.	Twechar bridge Hillhead bridge Erskine Ferry bridge	Opening bridge recommissioned Opening bridge recommissioned Opening bridge recommissioned	NEC/ECC , Option A, Design and construct
10b £1.75m	Bonnybridge Sylvania Way	New opening bridge Twin footbridges	NEC/ECC , Option A, Design and construct
10c	Leamington Lift bridge	Fix bridge in open position	TBD
11 £2.5m.	A80 Castlecary bridge	Fixed bridge Fixed bridge	NEC/ECC , Option A, Design and construct
12 £1.75m	Wyndford Auchinstarry	Fixed bridge Fixed bridge	NEC/ECC , Option A, Design and construct

Table 1. Major contracts

Note: In addition to the major contracts the Millennium Link project includes repairs to masonry bridges, aqueducts, weirs and culverts; installation of timber lock gates dredging and landscaping.

during the design and construction of The Millennium Link. However, the construction industry is used to such challenges with experienced people and organisations being available with established skills. The real early challenges lay in establishing the true role and value of the project and in putting together the required funding.

The improvement of canals in England and elsewhere had been shown to lead to wider regeneration with some notable successes such as in Birmingham[3]. The main drivers were the creation of colourful activity on the water and an improved environment along the canals. Activity could be quickly generated via the extensive network of canals across England.

The canals in Scotland are different. They are discrete and not integrated into a network. A significant level of activity on the water required at least the reopening of the canals between Glasgow and Edinburgh. By also reopening between the North Sea and the Atlantic, the west of Scotland sailing waters could be made accessible to an additional 500,000 craft in northern Europe[4]. Unlikely to risk the long open voyage to the more northerly and much larger Caledonian Canal, these coastal hopping smaller craft would provide a target market for the much more accessible Forth and Clyde Canal **[TABLE 2]**. Historic trade routes could be reopened, but this time for leisure craft.

Social inclusion maybe a current buzz phrase, but the underlying values are real. The canals pass through many areas of deprivation. Real sustainable development is now recognised as the combination of environmental, economic and social inclusion considerations in a cohesive way[5,6].

The Millennium Link project, which on the face of it, is about civil engineering is in fact the key first stage of an exemplar for sustainable development. It brings civil engineering into the mainstream. Civil engineers must be able to understand the real why of what they do.

NAVIGATION PARAMETER	**Forth & Clyde canal**	**Union canal**	**Caledonian canal**
Limiting Dimensions – Structures			
Length	20 .88m.	21.34m.	45.72 m.
Breadth	5.94m.	3.81m.	10.67 m.
Water Depth	1.83m.	1.06m.	4.11 m.
Height	3.0m.	2.74m.	27.4 m.
Canal channel			
Clear water depth over 6m wide channel	1.83m	1.06m	-
Towpath width	2.50m	1.80m	-
New bridges			
Height from water to canal bridge soffit	3.00m	2.74m	-
Height from towpath to bridge soffit	2.60m	2.45m	-
Clear water width	6.35m	3.85m	-
Clear water depth	1.83m	1.07m	-
Towpath width under fixed bridges	3.00m	2.50m	-

Table 2. Navigation parameters

PARTNERSHIP FUNDING

As the project was developed from the summer of 1994, a complex partnership and funding package had to be assembled. Restoration of navigation on the two canals required the raising of substantial capital funds. This investment had to be justified, but not necessarily with a direct commercial payback. The investors have their own objectives including social, economic and environmental aspects.

The revenue costs of the reopened canals also had to be justified. Step income had to match or exceed step costs to allow British Waterways to continue in the role of owner and manager.

British Waterways took the lead not only in engineering the restoration, but also in pulling together a partnership of agencies to create a funding plan and a bid to the Millennium Commission and bids for European Structural Funds.

Each funding partner supported the project because it met their own criteria, whether social, economic or environmental or indeed all three. All were united in a common desire to see the canals contribute once again to the life and economy of central Scotland.

The first bid was submitted to The Millennium Commission for funds from the National Lottery in April 1995 which was unsuccessful. The strength of the partnership and the strength of the basket of benefits which would flow from restoration led to a revised funding package and a revised bid to the Millennium Commission in November 1995.

A problem encountered in putting the whole scheme together was that both The Millennium Commission and the bodies disbursing European Regional Development Funding wished to be the last gap funders. It was not possible to secure the lottery funding without ERDF commitment, but ERDF could not legally be committed unless it was the final piece of funding required to allow a project to happen.

This was a common problem with lottery funded projects throughout the U.K. and eventually the funding bodies agreed to offer conditional awards. Essentially the whole funding package was put together on the basis that the contribution of each funder was conditional on all other funders committing.

The lead role and the financial risk lie with British Waterways. The overarching document is a contract between British Waterways and the Millennium Commission. This sets a completion date of September 2001 and defines project purpose as the restoration of navigation coast to coast and between Edinburgh and Glasgow. British Waterways has similar legal agreements with Scottish Enterprise and with the local authorities. All responsibility and risk for delivery of the £78million canal restoration including any cost overrun lies with British Waterways who are also the recipients of European Regional Development funding. This willingness by one party to take a strong lead and accept clear responsibility was instrumental in the success in securing the partnership and the funding package. Each funding partner has clarity of responsibility and contribution. A partnership Steering Committee chaired by a local authority councillor monitors progress **[TABLE 3]**.

FUNDER	AMOUNT
Millennium Commission	£32.2 m.
Scottish Enterprise Network	£18.7 m
Strathclyde European Partnership (ERDF)	£4.75 m.
Eastern Scotland European Partnership (ERDF)	£3.84 m.
British Waterways	£9.3 m.
City of Edinburgh Council	£1.7 m.
West Lothian Council	£0.55m.
Falkirk Council	£0.55 m.
North Lanarkshire Council	£1.28 m.
East Dunbartonshire Council	£0.44 m.
Glasgow City Council	£2.35 m.
West Dunbartonshire Council	£0.33 m.
Private and voluntary sector	£2.4 m.

Table 3. Project funders

FUNDRAISING – A CONTINUING THEME

The strength of the partnership and the value of the project were again demonstrated when late in 1997 an unexpected shortfall in funding became apparent. The partner agencies rallied and provided underwriting guarantees to cover the shortfall, avoiding the unravelling of the whole funding package, and allowing the project to proceed. Fundraising from additional sources became necessary and work is continuing with the objective of covering the shortfall to minimise the further call on the partners.

Fundraising will be a feature of the project well in to the future. The ongoing demands of further improvements to the canals themselves and their environment; the opportunities to link with community projects across the 110 kilometres; the opportunities for sustainable development on sites along the canals; the opportunities to link with educational initiatives; - all of these require ongoing work to secure funding packages.

MULTIPLE VALIDATIONS

Convincing partners of the value and benefits of The Millennium Link required multiple validations of the project or aspects of the project.

Local authorities were mostly interested in the local social and environmental benefits of the particular section of the project through their area. The support of canalside communities was as crucial as any projected benefits produced by professional analysis.

The Millennium Commission, having been convinced of the community support, appointed consulting engineers to carry out a detailed appraisal of technical and financial aspects including the likely achievement of the multitude of statutory approvals required.

Scottish Enterprise evaluated a broad range of economic and environmental benefits.

To obtain European Regional Development Funding required a full socio-economic cost benefit analysis in accordance with rules laid down by the European Commission. The analysis required the full cost to be justified regardless of the level of grant.

The built and natural heritage issues were carefully assessed. The canals are Scheduled Monuments. The entire project therefore had to be approved by Historic Scotland. The natural environmental aspects had to be approved by Scottish Natural Heritage to allow that body to validate the project to the satisfaction of the European Commission in Brussels.

Final funding details were confirmed in the early spring of 1998 and physical construction was inaugurated by Donald Dewar MP, the then Secretary of State for Scotland in March 1999.

THE WHEEL – A SYMBOLIC CHALLENGE

Historically, the two canals met at Falkirk, with a traditional flight of 11 locks overcoming the 110 ft. [33.5m.] level difference. A new route and a new method of connecting the canals have been devised intended as a landmark for modern engineering. Innovatively combining proven technologies, a striking rotating boat lift will transport leisure craft and visitors between the two canals. Its iconic design and its sheer scale and uniqueness should be a magnet for visitors and it is expected to become an international tourism attraction in its own right **[FIG. 2]**.

The Wheel is symbolic of the purpose of The Millennium Link. On the site of former opencast mining and an old tar works, it is also adjacent to the Antonine Wall, a precious ancient monument which marks the northernmost boundary of the Roman Empire. The Wheel is more than simply a device for moving boats between two canals. It utilises old industrial land, combines heritage with new tourism, but most importantly it is designed to reflect the new century. Its presence should stimulate development creating opportunities for local people, groups and businesses. The Wheel was conceived as a mechanism to stimulate sustainable change.

The Millennium Link is the first key stage of a long term sustainable development project. It utilises a redundant, albeit very important, part of Scotland's industrial heritage and by transforming it for new uses in a new century, the canals are planned to once again make an important contribution to the quality of life in central Scotland.

Figure 2. View of Falkirk Wheel. ***[British Waterways]***

THE ROLE OF CIVIL ENGINEERS

Throughout this project from the summer of 1994 until the funding package was finally secured in March 1998 and in to the construction phase the lead has been taken by civil engineers. Modern civil engineers, particularly new entrants, must recognise that their role goes well beyond a narrow perception of civil engineering. A full understanding of the context and purpose of physical projects is necessary. Understanding and respect for the natural and built environment and for the role of communities is important. Understanding of the mechanics of funding and government will allow engineers to be recognised as serious players.

Senior members of the profession know these issues only too well and must not be afraid to show a lead. Engineers are well suited to such a role, but can have a narrow focus. The early engineers were often close to scheme promoters. If engineers fully understand sustainability they will be well placed to rediscover that broad role.

References

1. Lindsay J., *The Canals of Scotland,* David & Charles, Newton Abbot, 1968.
2. *Forth and Clyde Canal Local Plan,* Strathclyde Regional Council, Glasgow, 1988.
3. *Canals – A Catalyst for Regeneration,* Conference Proc., British Waterways, Glasgow, 1995.
4. Ash Consultants and Coopers and Lybrand, *Tourism Study for British Waterways and Scottish Tourist Board*, British Waterways, Glasgow, 1995.
5. *Consultative Report – The Sustainable Development Project,* Eastern Scotland European Partnership, Dunfermline, 1998.
6. *Down To Earth – A Scottish Perspective on Sustainable Development,* The Scottish Office, Edinburgh, 1999.

The effect of site constraints on the design of the bridges for the Millennium Link

GEORGE C JOHNSTON
JMP Consultants Limited, Glasgow, UK.

INTRODUCTION

The 25 new road bridges constructed on the Millennium Link are small span, and in engineering terms are relatively simple designs so it would be imagined that their procurement, design and construction would be straightforward. However, very few of the structures are located on "green field sites" and almost all of them are on existing road corridors. Constraints to the construction choices were caused by the limitations on space, the existence of many major services and historical bridge foundations, and need to comply with today's road alignment standards..

BRIDGES ON THE CANALS - HISTORICAL BACKGROUND

The Forth & Clyde Canal was constructed between 1768 and 1790 to designs by John Smeaton and Richard Whitworth.

The original bridge crossings were of the opening bascule type with narrow timber decks on masonry abutments and wingwalls. Training walls comprised large timber balks placed in a triangular grillage pattern. A navigable width of 21-6" was provided, and with headroom unlimited, the canal developed into an important waterway for both industrial and pleasure shipping. After more than a century of use, many of the narrow timber bascule bridges were replaced by more robust steel swing bridges which had deep plate girder decks. In the closed position, these bridges provide very limited navigation clearance to the canal.

The bridges on the Edinburgh and Glasgow Union Canal which was completed in 1822, are almost entirely of masonry arch form, and this controls headroom. The majority of these bridges still exist and few new bridges have been constructed since the canal was completed in 1822..

The enacting of the "Forth and Clyde Canal (Extinguishment of Rights of Navigation) Act, 1962, removed the responsibility for maintaining navigation of the Forth and Clyde Canal and closed all 34miles 1,625 yards of canal to navigation. Reasons cited in the Act include "*the cost of providing, operating and maintaining and replacing opening bridges....greatly exceeds the like costs for fixed bridges.... and the provision of such fixed bridges and causeways would be consistent with the maintenance of the canal for purposes other than navigation.* The effect of the Act was to allow unfettered new construction of low-level roads and services across the canal. At some locations, it enabled the demolition of locks and the maintenance of the swing bridges as opening structures ceased. Closure of the Union Canal followed in 1965.

The Millennium Link. Thomas Telford, London, 2000, 15–22

The closures of the canals to navigation allowed canal infilling over extensive lengths and at Wester Hailes on the outskirts of Edinburgh and at Blairdardie and Clydebank to the west of Glasgow significant sections of canal were lost. The need to replace many of the narrow sub-standard canal crossings resulted in road authorities choosing to replace many old bridges with concrete culverts.

THE NEW BRIDGES

The aim of the Millennium Link project is to reverse the loss of the canal and in addition to the reconstruction of the infilled sections and the reconnection of the two canals at Falkirk, the project requires all obstructions to navigation to be removed. Culverted road crossings had to be replaced and this meant that 26 new road bridges and the re-alignment of adjacent approach roads had to be constructed. Locations and brief details are shown in figure 1 and table 1

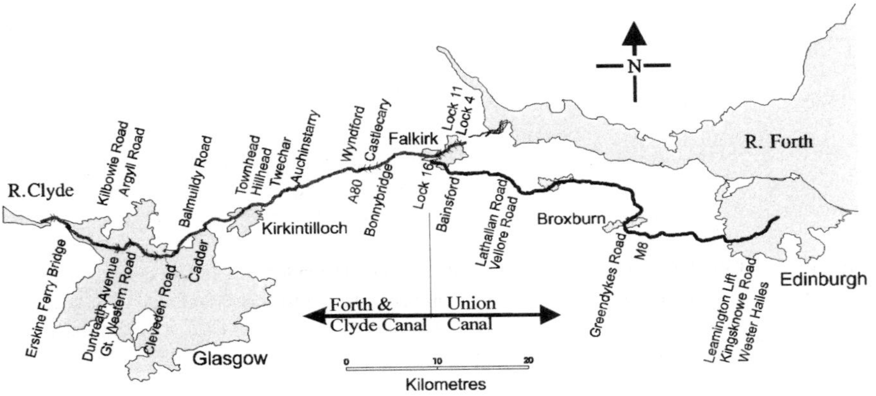

Figure 1. Location of Bridges.

The majority of the new bridges are located in urban or residential areas where construction space is limited and in a number of cases, they are in busy town centre locations where they are expected to be used heavily by road traffic and pedestrians. Three sites are on major trunk roads. Only six sites were in rural locations.

In addition to new “fixed” road bridges, existing bridges at Erskine Ferry, Hillhead, Twechar and Bonnybridge were refurbished or renewed as opening bridges.

The project was constructed using a number of Design and Build packages with Planning Consent being procured by British Waterways prior to the award of each contract. After award, the responsibility of purifying reserved matters or conditions of the Planning Consent fell to the successful contractor. In al contracts, the Design and Build Contractor was responsible for obtaining Construction Consent and Approval in Principle for the bridges.

To ensure the project programme could be achieved, British Waterways commissioned consultants to gather sufficient site information to enable Planning Consent to be obtained and to develop feasible construction options and contract documentation.

These consultants were also primarily responsible for preparing “exemplar designs” – effectively outline designs which were used to obtain Planning Consent at each site..

FORTH & CLYDE CANAL		
Bridge Name	**Road Authority**	**Notes**
Kilbowie Road	West Dunbartonshire Council	Major Road – Off line. Urban location
Argyll Road	West Dunbartonshire Council	Busy road serving commercial development. On line. Urban location
Dalmuir	West Dunbartonshire Council	New bridge and drop lock. Urban location
Duntreath Avenue	West Dunbartonshire Council & Glasgow City Council	Open Span Bridge – on line. Adjacent to Residential area.
Gt. Western Road & Clobberhill pipes	Glasgow City Council	Twin Open Span bridges. On line. Site of historical bascule bridge.
Clevedon Road	Glasgow City Council	On line. Residential area. Site of Historical Bascule Bridge
Balmiuldy Road	East Dunbartonshire Council	On line. Residential area. Site of historical bascule bridge.
Townhead, Kirkintilloch	East Dunbartonshire Council	Site of former bascule Bridge replaced by a 1930’s swing Bridge. Town center location.
Achinstarry	North Lanarkshire Council	Off line – rural location
Wyndford	North Lanarkshire Council	Off line –rural location
A80	The Scottish Executive	Major Dual carriage way –On line, Open span bridge, Rural location
Castlecarry	Falkirk Council	On line –rural location
Lock 16	Falkirk Council	Major road, Urban location
Camelon Road	Falkirk Council	Major Road.. Urban location
Bainsford Road	Falkirk Council	Important urban traffic distributer road – on line
Merer’s Bridge	Falkirk Council	Minor road serving industrial estate. Urban location
UNION CANAL		
Bridge Name	**Road Authority**	**Notes**
A801 Lathallan Road	Falkirk Council	Major Trunk Road- on line – Rural Location
Vellore Road	Falkirk Council	Minor country road- on line. Rural Location
Greendykes Road, Broxburn	West Lothian Council	Residential Area – on line.
M8 Crossing	The Scottish Executive	Motorway. 0n line. Rural Location.
Kingsknowe Road	Edinburgh Council	On line – Urban loocation
Wester Hailles	Edinburgh Council	Residential Area. Off line 4 road bridges and 3 footbridges.

Table 1- Fixed Bridge Sites

DEVELOPMENT OF BRIDGE FORM – EXEMPLAR DESIGNS

Generic Bridge Arrangements

The navigation envelope on the Forth & Clyde comprised a canal width of 6.5m and a towpath of 3m. Headroom to the canal was 3m. On the Union Canal the envelope was 4.5m canal and 2.5m towpath. Headroom was 2.74m. For the short spans needed to meet these requirements, reinforced concrete box or portal bridges provide an economical and practical solution as demonstrated by their use on recent canal crossings - on the Glasgow Canal Project and on the Union Canal at Preston Road in Linlithgow. In addition, preliminary studies had shown that, at most of the bridge locations, obtaining the minimum headroom would be extremely difficult.

Consultations with Local Authorities and Historic Scotland led to the early decision that the new fixed bridges should have architectural features and details which made them recognizable, by canal and road users, as a "family" of bridges associated with the Millennium Link.

A number of bridge design options were developed with architectural input from British Waterways. The "historical" bascule bridges on the Forth & Clyde canal, their swing bridge successors, and the arches used on the Union canal were totally different in form. It was felt that these differences should be reflected in the new bridges at the various locations.

Initially, attempts were made to replicate some of the features of the earlier bascule bridges by incorporating into the new bridges, variable depth decks, four rail parapets with twin central posts and a mock gearing feature on the pilasters. To reflect the original form of bridges on the Union Canal, the new bridges were detailed with arch features and reconstituted masonry facings.

However, during the design development process, these initial ideas were considered unsuitable and an alternative was developed. A simple bridge form enhanced by means of common "millennium link" features was accepted as being more appropriate. (figure 2).

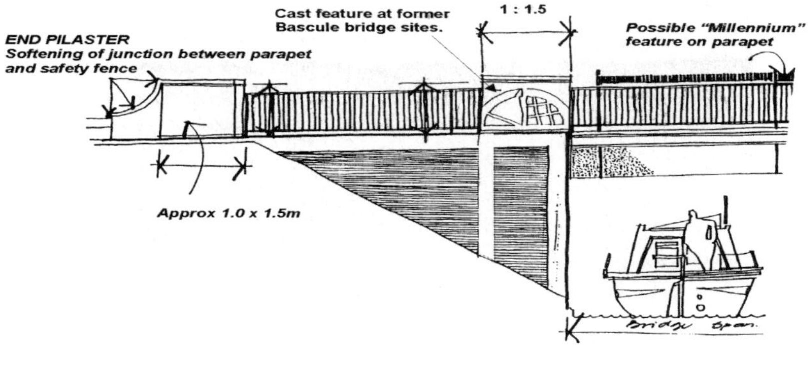

Figure 2.: "Millennium Link " Features.

At most of the sites of the new Union Canal bridges, there was insufficient headroom to provide a true arch bridge, but it was felt that the visual effect of an arch could be achieved by the use of a curved scalloped edge to the deck soffit.

In addition, because all existing bridges on the Union Canal are numbered, (from 1 at Edinburgh to 61 at Falkirk) it was decided to cast a relief of the bridge number in the deck fascia at the crown of the "arch".

A series of architectural sketches and details were prepared before the final proportions were finalized. In addition, photomontages of several sites were prepared to assist in the final choice, one of which is shown in figure 3.

The final features of the bridge "family" were:-

- The use of pilasters to help road users to relate the bridge to the Millennium Link and to create a smooth transition to the adjacent safety fences.
- The use of vertical infill to the bridge parapets where containment requirements allowed. (This was a compromise because the original bascule bridges had 4 rail parapets, whereas the swing bridges had vertical infill.) The incorporation of a continuous ...MMM...motif at the parapet top rail, both to identify with the millennium and to enhance appearance of the standard bridge parapets.
- The use of a common textured pattern for concrete surfaces which was to be used on all exposed faces of wingwalls and abutments.
- The incorporation of a feature to indicate the site of a former bascule bridge or where none previously existed, the new bridge name.
- The use of a heavy chamfer to deck edges (where structurally possible) to soften the impact of the structural depth necessary to incorporate services.
- A scalloped edge and bridge number for bridges on the Union Canal.

Figure 3 Photomontage of Vellore Road Bridge – Union Canal

SITE CONSTRAINTS

At that stage, it was recognized that the generic form would have to be adapted to suit the various constrains of navigation requirements, road design standards, the existence of statutory plant, presence of historical bridge foundations, and the need to acquire land.

Navigation Requirements and Engineering design criteria.

Although British Waterways set the navigation and canal works requirements, the engineering requirements comprising, bridge loading standards, parapet containment and verge/footway widths were prescribed by the several Local Road Authorities. In general they followed national standards. On all sites, the need to conform to the containment standards for bridge parapets meant that on most sites, a three rail parapet with mesh infill had to be adopted. In most cases, the bridge authorities considered the attachment of the MMMM feature on the parapet was an unacceptable departure from the standard and as this would prevent obtaining construction consent it resulted in the abandonment of attempting to incorporate the feature.

Road Design Standards

The acceptable new road alignment was agreed with the Road Authority. In many cases, the existing road geometry was already sub-standard and the need to raise the road to achieve navigation requirement caused, in some instances, an unacceptable deterioration of the standard. In several locations minor relaxations to the standards were accepted but at some sites, the conflicting requirements of road profile and canal headroom could only be achieved by relocating locks to lower the canal or by providing new locks and bridges – Camelon Road and Bainsford Road in Falkirk being typical examples. At Dalmuir near Clydebank the only practical solution was the provision of a drop lock and a new bridge.

At a number of sites, the difficulties with road geometry was so severe that the idea of a fixed bridge had to be abandoned. This occurred at the Hillhead, Kirkintilloch where the canal was constructed on sidelong ground and the existing swing bridge was located on a road with a 7% road gradient. The fixed bridge needed a road level 2m higher and this was not practicable.

A equally difficult site was at Twechar near Kilsyth where the 1960's bascule bridge was located at the bottom of a steep gradient and immediately adjacent to the B802/B8023 "T" junction.

Although fixed bridge solutions had been developed for both of these sites, they were less than practicable and not favoured by the Local Authority. The final proposals reverted to refurbishment of the opening bridges.

At three sites, where headroom permitted, Great Western Road, Duntreath Avenue and A80, an open span form of structure was preferred, but detailing consistent with the closed span bridges was used. (figure 4)

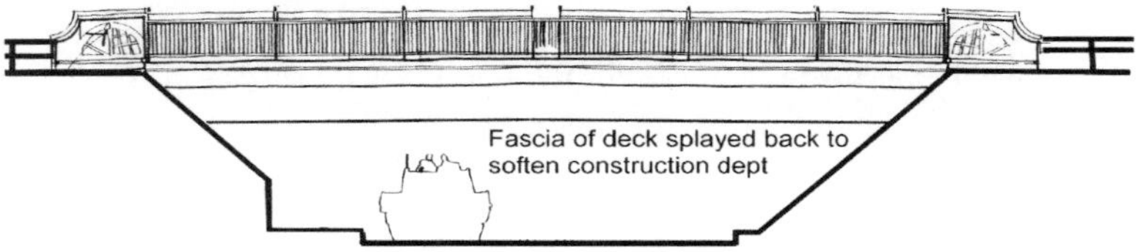

Figure 4 Open Span Bridge

The Duntreath Avenue bridge carries a wide traffic distributer road on a high embankment adjacent to lock 36. The new bridge location was close to the junction of Duntreath Avenue and the westbound slip road to the A82 dual carriageway. Moving the bridge was not feasible because of adjacent housing. These constraints resulted in the bridge being off the line of the lock and this introduced a "chicane" into the canal between the lock and the bridge hole. An open span bridge was used to give better visibility to canal traffic. The bridge, which had piled foundations, was constructed in two stages while maintaining traffic light controlled one way working on Duntreath Avenue.

At Great Western Road a wide structure was needed to carry the dual carriageway and two separate bridges of open span form with a wide uncovered "light well " located in the road's central reserve was adopted. This had the added advantage that the abutments of the historical lifting bridge would be exposed and this would form an interesting historical feature.

The existence of statutory undertakers plant.
Statutory Undertakers apparatus was abundant at almost all new bridge locations and on may sites, the design and cost of the canal crossing was adversely affected by the need to construct bridges deeper or wider than was structurally necessary. Many of the bridges had to be constructed in phases to allow apparatus to be diverted and protected at the relevant time while maintaining traffic flow.

At A801 Lathallan Road, a high speed trunk road linking the M8 to the M9, the bridge had to accommodate a 760 diameter water main. The A801 road alignment comprised a 3% gradient falling north to the M9. Headroom to the canal was extremely limited and any appreciable raising of the road profile meant that a long length of realignment was needed to achieve the acceptable vertical curvature. Raising the road alone was not acceptable as the increased road make-up increased the loading on an adjacent railway underbridge. The "exemplar" solution was to move the canal "uphill" and use minimum construction depth for the bridge deck. This necessitated the adoption of an above ground location for the diverted watermain. (figure 5)

Figure 5. Exemplar Design solution at A801 Lathallan Road.

All sites had difficulties with services, the most notable being M8 (750 diameter HP gas), Wester Hailles Road (900mm diameter sewer) and Cleveden Road, Glasgow where a 750mm diameter medium pressure gas main had to be accommodated and the bridge had to be constructed in three stages while the existing road had to be kept open to traffic signal controlled one way traffic at all times. This was a particularly difficult site because there was no scope to change from the existing road geometry and the required headroom was just achievable.

The remains of historical structures or foundations buried in embankments
As the majority of bridges had to be constructed on the line of the existing road corridor it was expected that – particularly on the Forth & Clyde - the historical remains of old bridges would be found and it was hoped that some of these remains could be incorporated into the new structures. This desire to utilize the foundations of historical structures meant that deviations from the generic bridge form would be required.

The town centre site at Townhead, Kirkintilloch is a particular example where the foundations of the 1930's swing bridge (a site of a previous bascule bridge) were known to exist and drawings were available. While the reuse of the massive concrete foundations meant that cost savings would be achievable, the historical remains were too narrow to accommodate the required road width and a complicated structural arrangement was needed.

The new deck needed to overhang the edges of the abutments and the construction depth had to be much greater than the minimum in order to accommodate an existing bank of telecommunications cables. Figure 6.

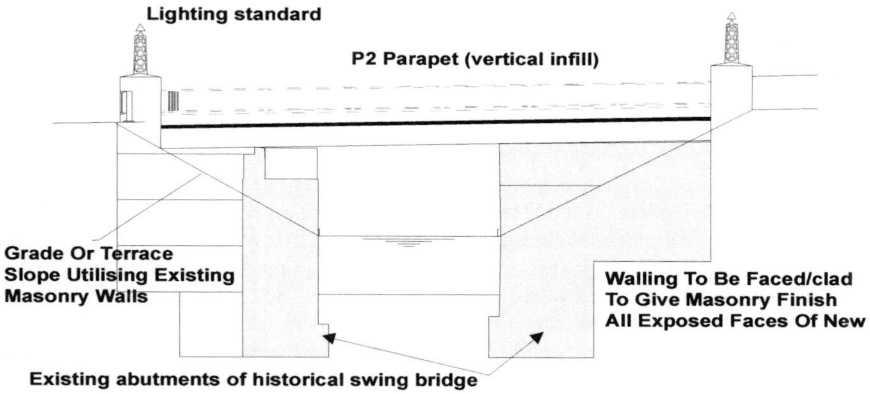

Figure 6. Townhead Bridge, Kirkintilloch

The need to acquire land.
Although British waterways had retained ownership of the historical canal solum and a limited strip of land either side, they had to acquire additional parcels of land to accommodate realigned roads and canal diversions. All land was acquired by British Waterways by agreement and without using compulsory purchase. Negotiations for land inevitably had an effect on the available options for individual sites but had little effect on the bridge design.

Although much of the land needed to accommodate the raised road corridors was owned by the relevant Local Authority, at two sites, M8 and Balmuildy Road, Bishopbriggs significant areas of privately owned land was required.

At Balmuildy Road, the initial exemplar design comprised an off line solution which required substantial land take to accommodate the new bridge and approach roads. However, concerns over possible difficulties with land acquisition led to the constructed on line solution despite the added traffic management and construction difficulties of building on top of an existing bridge that this caused. At the M8, the motorway was on a 4% gradient at the site of the canal and the only solution was to reconstruct the canal on a different line through farmland to a point where more headroom could be achieved. This point coincided with the location of a 750 diameter strategic gas main crossing under the motorway. With the cooperation of the landowner and Transco, extensive pre-construction surveys were carried out to determine the exact location of the main and a methodology for working adjacent to the main were agreed. This allowed the new bridge to be constructed adjacent to the main with a saving of £300,000 to the project. The bridge was constructed with 19mm headroom to spare.

ACKNOWLEDGEMENTS

The author acknowledges with thanks the assistance received from his colleagues in the preparation of this paper.

Aqueduct construction in the canal age – a case study

MRS SANDRA PURVES, BSc, CEng, MICE, FSA(Scot), ICE Panel for Historical Engineering Works, Secretary Scottish Group

INTRODUCTION

Various routes were proposed to link Edinburgh to the west by canal following the opening of the Forth & Clyde Canal in 1790. The route eventually chosen was that proposed by Hugh Baird, the Company engineer for the Forth & Clyde Canal.[1] His proposal was for a contour canal to run from "*the Lothian Road near the City Edinburgh to join the Forth & Clyde Canal at Falkirk in the county of Stirling*." [2]

In his report of 1813 on the proposed Edinburgh and Glasgow Union Canal Baird[3] identified several engineering structures which would be required. The first was a series of nine locks to take the canal from Lock 16 on the Forth & Clyde Canal to the level of the canal itself (eleven were built). The others were aqueducts over the river Avon, the river Almond and the Water of Leith at Slateford. The estimate of costs for these aqueducts all included the cost of puddling. However he discusses the idea of using cast iron. " *Some very extensive Aqueducts have lately been erected in England of cast iron; they are said to cost less than of stone. If they are on enquiry found to be less expensive, and as sufficient as of stone, they may be adopted, as it will be a great saving of time and execution of the work.*"[4]

The aqueduct over the Avon was expected to be expensive due to its size and the cost of getting materials to the site. On the Water of Leith crossing he remarks "*This I propose crossing by an Aqueduct of 4 arches of 50 feet each, the height being 43 feet; as the grounds on both banks consist of a kind of pleasure ground it will be necessary to have this Bridge neatly finished and the lands properly fenced.*"[5]

DEVELOPMENT

The first major canal aqueduct to be built in Britain was Brindley's three-arched aqueduct at Barton, which from 1771carried the Bridgewater canal over the river Irwell[6] (replaced in 1893 by the Barton swing aqueduct). It was considered an outstanding feet of construction for its time and gave Brindley an enthusiasm for building aqueducts. This and subsequent structures were masonry with a puddled lining to keep the aqueduct watertight. This type of aqueduct tended to be massive and often prone to leaking and not always structurally stable.

Other engineers however produced some successful masonry aqueducts; Smeaton's Luggie aqueduct[7] and Whitworth's Kelvin aqueduct[8] on the Forth and Clyde canal and Rennie's Lune aqueduct[9] on the Lancaster canal being three such examples. To be successful these structures required large quantities of masonry; this could be quite costly and take time to build. However between 1776 & 1779 a new construction material had been used to build a bridge over the river Seven near Coalbrook Dale. This material was cast iron.

The Millennium Link. Thomas Telford, London, 2000, 23–31

TELFORD'S INVOLVEMENT

Thomas Telford was one of the first engineers to exploit the use of cast iron for bridges and in the building of aqueducts, where he used it to form the channel for the water to replace the clay puddle. These included Longdon on Tern,[10] Chirk (cast iron base plate, with the sides of the waterway made waterproof using ashlar masonry and hard burnt bricks in Parker's cement)[11] and Pontcysyllte.[12] On the Grand Junction Canal when Jessop's Cosgrove masonry aqueduct [13] over the Great Ouse at Wolverton collapsed it was rebuilt in cast iron on a masonry pier and with brick abutments.[14]

In 1815 the Canal Company asked Telford to comment on Baird's line for the Union Canal. His report was favourable.[15] Telford was employed as Consultant or Principal Engineer and acted for the Canal Company in ensuring a smooth path for the "Canal Bill" through Parliament. Baird was appointed Resident Engineer in August 1817.[16]

THE GREAT AQUEDUCTS ON THE UNION CANAL

Work on the construction of the Union Canal started with a ceremonial dig on Tuesday 3rd March 1818, which took place after the General Assembly of the Proprietors of the Canal Company. A plan of the proposal for the iron aqueduct at Slateford was exhibited here.[17]

The work on the canal was divided up into lots, the large aqueducts at Slateford and across the River Avon and some medium sized ones including the aqueduct over the river Almond being individual lots. The culverts and smaller aqueducts were included in the lots for the cutting of the line of the canal.

In order to get the job underway the work was advertised as soon as Hugh Baird had plans and specifications ready. Although the work was to start at the east end and progress westwards the aqueduct at Slateford was not included in with the parcel of lots for the first 10 miles which was advertised on Thursday, December 18th 1817.[18] The second portion of work, the thirteen miles of canal from the river Almond to the river Avon was advertised on Monday March 2nd 1818, including the masonry for the aqueduct at Slateford and the aqueduct over the river Avon.[19]

On 17th March Baird wrote to Telford referring to "*The Iron Aqueduct at Slateford, which I have made much like your own perhaps fully as strong- I have advised delaying the advertisement for the Iron Work, wishing before doing it to consult you on the construction and this I'll communicate to the Committee.*"[20] The Iron aqueduct was attracting the attention of potential contractors. Two day before Baird had written to Telford he had received a visit from James Thomson, who worked for Hazledine, the Iron founder who had constructed the Pontcysyllte aqueduct. According to Baird "*He (Mr Thomson) suggested an Iron House in the West – who has offered to supply if he becomes Contractor.*"[21] Baird suggested that Mr Thomson produced a section and elevation of the proposed ironwork and take it with him to London to show Mr Telford.

The committee however was having doubts about buying plates which would be manufactured at a great distance from the works rather than locally.[22] To transport the iron from Shropshire it would have to be sent by boat down the river Seven, round the south coast of England, through the English Channel, and up to Leith.[23] Copies of the specifications for the masonry and ironwork for Slateford Aqueduct were sent to Telford by Baird on 26th March for his consideration. The question of whether the contract should be for "*the Castings delivered on the Ground or erected and finished*"[24] was also raised.

On 3rd April the Canal Company Committee met to consider the tenders for the aqueducts. The lowest overall tender for the aqueducts was from Hugh McIntosh, the Contractor for the eastern end of the Canal and that for the stonework of Slateford Messrs. Craven, Whittaker & Nowell. Since discussions were on going with Telford on the design it was decided to not to let the contract for the Avon aqueduct. After a gentleman's agreement, between the interested parties, the committee decided to award the work at Slateford to McIntosh as it was situated in the middle of the section they were already engaged on.[25]

Meanwhile Mr Thomson was busy producing drawings of ironwork for the aqueduct at Slateford under Telford's direction. He also produced a break down of the weight of cast ironwork that would be required in one arch. This resulted on 30th April 1818 in an estimate being sent to Telford, which was signed by W. Hazledine & James Thomson
"*An Estimate to furnish the cast and wrought iron work of an aqueduct to be erected at Slateford for the Edinburgh and Glasgow Union Canal ------------ The Contractors to find scaffolding with all the other materials to erect and complete the whole agreeable to drawings & Specifications (except the materials for jointing the castings watertight)*
The whole castings to be from best No 2 Shropshire Iron @ fourteen pounds five shillings per ton £14-5-0 and all the wrought Ironwork to be made from best charcoal iron @ sixpence per lb. say £0-0-6."[26]

The result of this was that Baird produced revised specifications for the two aqueducts[27], Advertisements were placed in the Edinburgh, Glasgow and London papers on July 18th 1818 for the following contracts; the construction of a cast iron aqueduct at Slateford and a stone aqueduct over the river Avon valley near Manuel Mill.[28] These tenders were opened on 11th August, the minute book recording "*Thereafter the tenders for the great Aqueducts at Slateford and at Avon River were opened by the committee when the offers of Messrs. Craven, Whitaker & Nowell appearing to the Committee to be by far the most eligible they accepted the same for both these bridges.*"[29]

Unfortunately the details of the other interested parties are not recorded. The committee must have decided that it was no longer appropriate to award it to Hugh McIntosh and either the cost and problems of transportation or the desire to obtain the iron work locally ruled out Hazledine.

CONTRACTORS

Contracting firms, which had evolved in Scotland on the Forth & Clyde canal since 1768, had developed somewhat towards the end of the canal age (about 1820). All prospective contractors had to put up security and details of previous experience. Several contractors who had previously carried out work on their own account began at this time to join forces with others in order to raise sufficient funds for such work. Messrs Craven, Whitaker & Nowell had joined together to construct a masonry bridge over the river Ouse at York prior to tendering for work on the Union Canal. See Fig.1.

Their expertise was in stonework, which was fine for constructing the pillars for the aqueducts as by utilising cross wall the quantity of stone required, and thus cost, was reduced. The standard of workmanship needed was however greater, the subject of which led to disagreement between Baird and the overseers on one hand and the contractor on several occasions.[30]

Messrs. Craven, Whitaker & Nowell

"Builders & Contractors for Public Works at Keighley, Bingley, & Dewsbury, Yorks."
The three partners in the firm were; Hiram Craven (1780-1842), Samuel Whitaker (c1783 - 1821) & Joseph Nowell (1784-1836).

Two of the partners appear to have started out as stone masons or stone merchants or were from families who were. These were Hiram Craven and Joseph Nowell who had both undertaken bridge work on their own account previously. Samuel Whitaker, who was unfortunately killed in an accident to the London & Edinburgh Mail Coach whilst travelling home from the Canal, was described as an architect. Hiram Craven's son John (1800-1870) married Frances Nowell, Joseph's younger sister.

The partnership's first project was the building of a masonry bridge over the River Ouse at York. It was on the strength of this job that they tendered for the aqueducts on the Union Canal. The day to day supervision of the aqueducts was the responsibility of Joseph Nowell, the junior partner and John Craven. Although the firm was dissolved after Samuel Whitaker's death, the remaining partners and John Craven continued successfully as Contractors.

Fig.1 Notes on the firm of Messrs. Craven, Whitaker & Nowell
Sources
Frank D. Smith Files, FSB052, Institution of Civil Engineers Library
John Craven's obituary, York Herald, July 1870
The Nowells – A Remarkable Dewsbury Family – originally printed in the Dewsbury Reporter, July 20th 1932

CONTRACT VARIATIONS

Messrs Craven, Whitaker & Nowell's tender for the Avon aqueduct contained an attached alteration for its construction.[31] Although details of this alteration do not appear to exist the proposal was basically for the use of a cast iron trough rather than clay puddle for the water channel. This resulted in Baird and Telford conferring on the design and eventually agreeing that both aqueducts would be masonry with a cast iron trough. On Dec 17 1818 the Committee of Management wrote to Messrs Craven, Whitaker & Nowell agreeing to the alterations, despite an increase in cost of £1000.[32] Fig 2.

Work on Slateford aqueduct (Fig.3) had commenced by March 1819, the minute book recording on 1st March "*Mr Burke* (Assistant Engineer) *having mentioned that the foundation of Slateford Aqueduct was laid yesterday and that it was customary to allow something to the workmen the meeting ordered £5 to be given to them.*"[33] The committee was keen to make sure the work was done well and speedily as on 19th March they instructed Hugh Baird "*to order two Mortar machines one for Slateford and Avon aqueducts value of each £36 and to look out for proper overseers of Mason work on trial.*"[34]

ALMOND AQUEDUCT

Originally the aqueduct over the river Almond was to have been one large arch and embankment, with a separate aqueduct over the "Linn Mill" road. In June Baird altered the plans to a bridge of 5 arches of the same construction as the other two major aqueducts,[35] (Figs. 4 & 5) by doing this he anticipated saving money on the embankment, puddling and Linn Mill aqueduct. The contract was let to Charles Stewart who started work in about

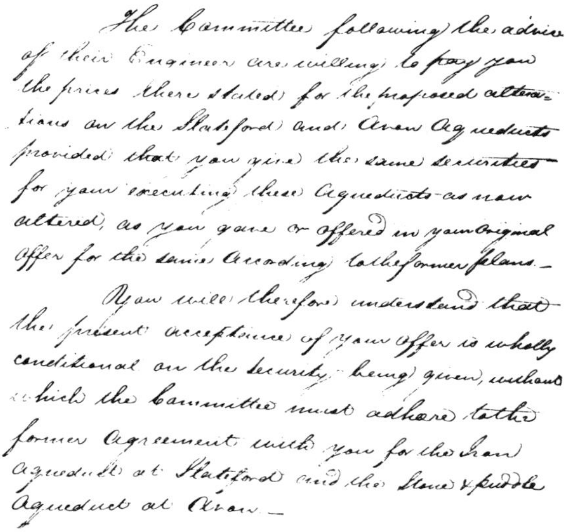

The Committee following the advice of their Engineer are willing to pay you the prices there stated for the proposed alterations on the Slateford and Avon Aqueducts provided that you give the same securities for your executing these Aqueducts as now altered, as you gave or offered in your Original Offer for the same According to the former plans.—

You will therefore understand that the present acceptance of your Offer is wholly conditional on the security being given, without which the Committee must adhere to the former Agreement with you for the Iron Aqueduct at Slateford and the Stone & puddle Aqueduct at Avon.—

Fig. 2. Extract from Letter to Messrs. Craven Whitaker & Nowell agreeing to their proposed alteration to the almond aqueduct from G. Moncreiff, 17th Dec 1818 [32]

Fig.3 Slateford Aqueduct in 1823, (Edinburgh City Libraries)

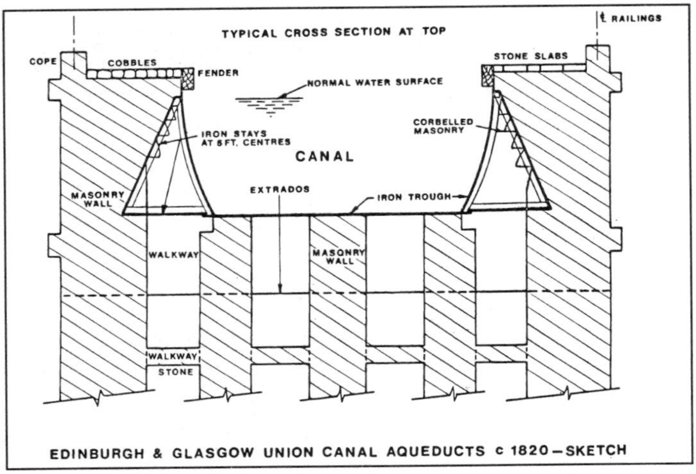

Fig. 4 Cross Section showing cast iron trough.

CAST-IRON WORK

Water-Way.

The Cast-Iron Water-way will extend 41 feet over the Abutment face at each end, in all 372 feet long, to be 10 feet 8 inches wide at bottom, 13 feet 6 inches wide at top, and 6 feet 10 inches deep. The Bottom Plates will decline to 3 feet below bottom at each end; and the Sides must widen out with the Wings to 26 feet wide at the extremity. The Bottom Plates will be 7 feet 8 inches by 5 feet, and 4 feet by 5 feet; to be three-fourths of an inch thick, with their Flanges one inch thick. The Side Plates will be 6 feet 11 inches high, 5 feet broad, full three-fourths of an inch thick, with a Flange of 3 inches broad, and one inch thick all round, and a Rib up the middle of them; the Plates to be curves, and otherwise constructed, as particularly shewn in the Section to a large scale. All the Flanges to be properly constructed for packing the joints off the inside, and to have a Bolt of seven-eighths of an inch diameter at every 9 inches of the Flanges. The Side Tyes, of which there must be one at every 5 feet, to be 4 feet long, 9 inches broad, by 1 inch thick, properly joggled into the Soal Plates and Stone String Course, having a Rib of 2 1/2 inches by three-fourths of an inch. On each Side Tye there must be a stay of 6 feet long, 6 inches broad and three-fourths of an inch thick, with a Rib of 4 inches, by three-fourths of an inch thick, properly fitted with Brackets, and bolted to the Side-Plates and Tyes. The Castings to be made of good No. 2 Iron, either from the Air furnace or Cupelo. The Joints to be made perfectly water-tight, by being caulked with white rope yarn soaked in Mac-adam's Mineral Paint, or other such substance as shall be approven of by the Engineer.

Fig.5. Details for Cast Iron Trough for Almond Aqueduct – Extract from Specification No 12 [46]

August 1819. By the end of October concern was being expressed regarding Stewart's ability to carry out this work and by 16th November the committee was looking at legally ending it.[36] Offers were then sought from Messrs Craven, Whitaker & Nowell and Messrs Hughes & Williams for the contract. Much deliberation followed over which offer to accept, "*from the experience the Committee had already had of both the present offers they could not help having more confidence in Craven Whitaker & Co as being regular Bridge builders and being in much more established and respectable circumstances than the others*". After discussions between the two potential contractors the later withdrew their offer and the contract was awarded to Messrs Craven, Whitaker & Nowell.[37]

CONSTRUCTION

Baird and Joseph Nowell failed to agree at times, Nowell complaining about Baird harassing the Contractor, whilst Baird complained that Nowell was constantly telling the men that they were doing their work too well.[38] Certainly the minutes and letter books keep referring to points where the contractors were at variance with the specifications, e.g. using undersized or unsuitable stones etc.[39] The contractor was also unable or unwilling to find sufficient work force; there are several complaints about lack of progress. Erection of the ironwork for the troughs also gave them some trouble, due to lack of experience.

Whether it was George Moncreiff, the Canal Company Clerk's, decree that the ironwork should be obtained locally or the cost of transportation that influenced matters is not recorded but it was supplied by Mr Anderson[40] of the Leith Walk Foundry, Leith[41]. The detailing for the plates was done by Robert Baird,[42] Hugh's younger brother who was joint partner with Hugh in the firm of "*Hugh & Robert Baird, Founders and Millwrights at the Canal Bason, Hamiltonhill, near Glasgow*"[43]. It was in the end fortunate that the supply was local as the contractor found he was short of some plates on the Avon aqueduct.[44]

The pressure to get the work completed resulted in disagreement between the contractor and engineer. The contractor wanted to fill the trough with water the minute the tough was complete, whilst the engineer was only prepared to allow the trough to be filled to a depth of 2 to 3 feet until such time as the masonry water walls to the trough were satisfactorily completed.[45]

ENGINEERING CONSIDERATIONS

Baird as resident engineer was responsible for the design aspects of the project. Whilst the records refer to plans and elevations he produced these do not appear to still exist, however copies of his Specifications Nos. 1-12 have survived[46].

At the time these aqueducts were being constructed the science of 'strength of materials' was very much in its infancy. Tredgold's book on cast iron, probably the first textbook to attempt to deal solely with the subject, was not published until 1822.[47] Baird was aware of at least some of the properties of cast iron being concerned about the effect of too great a pressure on the side plates and of impact if the trough was filled to its full depth before the masonry walls were built. Perhaps the best indication we have of Baird's knowledge of the capabilities of cast iron is reflected in the aqueduct which carries the feeder over the river Almond for which unfortunately no specifications or other details have been located.

Telford, although he was involved in the discussions on and approved the final design, still questioned in his article on Inland Navigation " *how far it was judicious to incur the expense of providing two modes of security in the same structure* "[48]

Bores were sunk along the line of such structures as the aqueducts, and rock levels indicated on the plans. This was the Achilles heel of the project. No allowance seems to have been made for possible variation of rock head levels, which in several cases proved to be at greater depth than anticipated.[49] This was one of the main reasons the cost of construction was significantly higher than Baird's estimate. A substantial claim was submitted for these variations by Messrs. Craven Whitaker & Nowell. This claim was put to arbitration with James Walker, Civil Engineer, London being the arbiter.[50]

CONCLUSIONS

Considering the lack of construction equipment and the isolation of the sites, in particular the Avon valley, these *magnificent* aqueducts are a tribute to the skills and abilities of the engineers and contractors who built them.

The *GREAT AQUEDUCTS* were state of the art in their day but did not influence the design of later canals. The strength of the Union Canal lies in its aqueducts with their standardised arches which enabled the canal to be made level over 31 miles Besides those built originally three new ones in reinforced concrete were built during the twentieth century. The first, the 'Prince Charlie Aqueduct' was to allow the Lanark road to be widened and replaced the original in 1937. The second at Bonnington in 1976 and lastly the aqueduct over the City By-pass at Hermiston built in 1989. The faith in the canal's future shown in building this latter structure was one of factors that eventually allowed the Millennium Link Project to happen. Significantly the main feature of the Millennium Link Project to be constructed in connection with the Union Canal will be the twenty-first century aqueduct on the extension to the canal leading up to the 'Falkirk Wheel'.

REFERENCES

---

[1] *State of Facts & Observations relative to the Affairs of The Forth & Clyde Navigation,* January 1816, p17 (Forth & Clyde and Union Canals, Mitchell Library, Glasgow Room)
[2] National Archives of Scotland, BR/EGU/4/6/9
[3] Baird, Hugh, Report on the proposed Edinburgh and Glasgow Union Canal 1813, (Edinburgh City Library)
[4] Ibid. p 6
[5] Ibid. p 7
[6] McKnight, Hugh, *The Shell Book of Inland Waterways,* David & Charles, Newton Abbot, 1975, p67
[7] Paxton, R. A., Paper 12216, Proceedings of ICE , Civil Engineering, May 2000. p 88
[8] Paxton, R. A., Paper 12216, Proceedings of ICE , Civil Engineering, May 2000. p 88
[9] Unknown, *Aqueduct,* Encyclopædia Britannica, Edinburgh, Seventh Edition, 1842, p332
[10] Cragg, R,. Civil Engineering Heritage, Wales & West Central England. Thomas Telford, London, 1997, p252
[11] Ibid, p 48-9
[12] Ibid. p 46-7
[13] Labrum, E. A. *Civil Engineering Heritage, Eastern & Central England.* Thomas Telford, London, 1994, pp236-7
[14] Faulkner, Alan H., *The Wolverton Aqueduct: A Case Study in Canal Engineering*, Transport History, Volume 2 Number 2 July 1969, David & Charles, Newton Abbot, p164
[15] Telford, T, *Mr Telford's Report on the Union Canal*, The Scots Magazine, April 1815, pp337-9
[16] Lindsay, J. *The Canals of Scotland*, David & Charles, Newton Abbot, 1968, p 37
[17] *Union Canal Commencement*, Scotsman No.59, Saturday March 7, 1818, p77
[18] Advertisement, Thursday, December 18, 1817, in the Edinburgh Evening Courant
[19] Advertisement, Thursday, Monday, March 2, 1818, in the Edinburgh Evening Courant
[20] Baird, Hugh, *Letter to Telford, dated 17th March 1818,* Telford mss, Edinburgh and Glasgow Union Canal, ICE archives, p234
[21] Ibid.p234
[22] Moncrieff, George, *Letter to Telford,* Telford mss, Op. Cit., p236

[23] Telford , T, *Letter to George Moncrieff,,* Telford mss, Op. Cit., p244
[24] Baird, Hugh, *Letter to Telford, dated 26th March 1818,* Op. Cit., p280
[25] National Archives of Scotland, BR/EGU/1/2, Minute Book, 3 April, 1818
[26] Hazledine, W & Thomsom, J., *Letter to Telford, dated 30th April 1818,* Telford mss, Op. Cit., p184
[27] NAS, BR/EGU/3/1, Specifications, 1818
[28] Advertisement, July 18, 1818, in the Edinburgh Evening Courant,
[29] NAS, BR/EGU/1/2, Minute Book, 11 August 1818
[30] NAS, BR/EGU/4/6/5, Letter Book
[31] Ibid. 12 August 1818, p1
[32] Ibid. 17 December 1818, letter No. 6, p5
[33] NAS, BR/EGU/1/2, 1st March 1819
[34] Ibid. 19th March 1819
[35] Ibid. 1st June 1819
[36] Ibid. 16th November 1819
[37] Ibid. 1st December 1819
[38] NAS BR/EGU/4/6/5, p70
[39] NAS BR/EGU/4/6/5 & NAS, BR/EGU/1/2
[40] Baird, H., *Letter to Messrs. Craven, Whitaker, & Nowell,* NAS BR/EGU/4/6/5, 17th January 1821, p114
[41] The Post Office Annual Directory for 1817-18, Leith.
[42] Baird, H., *Letter to Messrs. Craven, Whitaker, & Nowell,* NAS BR/EGU/4/6/5, 27th March 1821, p125
[43] Post Office Glasgow Directory, 1799
[44] Baird, H., *Letter to Messrs. Craven, Whitaker, & Nowell,* NAS BR/EGU/4/6/5, 17th January 1821, p114
[45] Baird, H., *Letter to Messrs. Craven, Whitaker, & Nowell,* NAS BR/EGU/4/6/5, 13th January 1821, p103
[46] NAS, BR/EGU/3/1, Specifications, Nos. 1-11, British Waterways Archive, Glasgow, No.12
[47] Tredgold, T., Practical Essay on the Strength of Cast Iron, 1822
[48] Telford, T. Navigation Inland, Edinburgh Encyclopædia, Edinburgh, 1830. XV, 311, First published in 1821.
[49] BR/EGU/4/6/5, Memorial for Craven, Whitaker & Nowell, p5
[50] Ibid. p1

Recording canal structures for the archive

GRAHAM J. DOUGLAS

Both the Union and Forth and Clyde canals are Scheduled Ancient Monuments and the Scheduling covers the canals and all structures along their length whether dating from their constuction, 1770s for the Forth and Clyde canal and 1820s for the Union canal, and right up to the 1960s, when they were closed to traffic and partly infilled. Part of the consent conditions from Historic Scotland to allow British Waterawys to develop the canals, was for an archaeological adviser to be appointed to maintain a watching brief and to ensure the proper recording of any structure revealed, altered or demolished during the works.

The archaeological adviser is appointed subject to the approval of Historic Scotland and is resposible for providing advice to the contractor and client for each site where the new work crosses, incorporates or disturbes the old canals and their assosiated structures.

The purpose of this article is to describe the work of the archaeological adviser for this project, which has become known as The Millennium Link. The work is largely concerned with graphically recording the scheduled site. In the end, what has been achived is a set of records of sites which no longer exist, the records being of varying level of detail, showing construction methods and materials used.

Part ot the Schelduled Ancient Monument Concent is that the records compiled are deposited at the National Monuments Record of Scotland. This is a public record office in Edinburgh which is the main archive for the built heritage in Scotland.

The first part of the process starts with a contract between the archaeological adviser and the contractor who is carrying out the work on behalf of British Waterways. The adviser is provided with drawings for the structure to be built, along with copies of any plans or drawings of earlier structures on the site plus any archival material or photographs. A good deal of the archival and documentry research had therefore already been carried out by the design consultants before the adviser was appointed. The main role for the archaeological adviser was therefore one of a field recorder rather than archival researcher. Provision of an archaeological watching breif was part of the requirement for Schelduled Ancient Monument concent. To maintain an effective watching brief on all the works, close liasion with each site agent was essential. Most of the agents had never worked with an archaeologist before and ensuring they understood what the archaeologist was required to do and why was the key to sucess in this job.

At the outset of the project several sites were noted as being potentially very

The Millennium Link. Thomas Telford, London, 2000, 32–37

significant, for example several bridges and locks. Because these structures were to be demolished, a very detailed account of their construction was possible, much more so then if they had remained standing. No detailed construction or design drawings have survived so recording the structures during their demolition gives the best oppotunity to understand them. Detailed field recording plays a large role in understanding industrial monuments such as this.

The records of each site or feature followed the same format and varied in detail according to the importance, complexity and accessibility of the site. Recording methods used were photographs, dimensioned annotated site sketches, scale drawings, rubbings of incised or carved detail and written descriptions. Comparative drawings were also used for recurring features such as milestones, masonry bridges and mason's marks to highlight any shared characteristics. Often what these drwaings did was to show how certain types of structures which everyone had assumed to be the same were in fact different, no two milestones were exactly alike.

The practicality of carrying out detailed survey work on a large scale construction site such as this did mean that conditions were often not ideal. Modern civil engineering practices do not stop for rain, hail or snow, but these conditions are not kind to the archaeologist with the required hands on approach to detailed recording. Time constraints are always present, demolition time using modern machinery is surprisingly quick and does not lead to delicate dismantling. Health and safety practices must be observed, and the contaminated nature of the canal in places restricted access. The work of recording is still going on, but to date 18 sites on the Union Canal, and 57 on the Forth and Clyde have been recorded. A break down of the 18 Union canal sites is thus; 6 bridges of the 1960s period built using concrete pipes, 1 of the same period using a corrugated steel pipe, 2 timber safety gates, 1 stable building, 7 masonry bridges and the remains of a lime kiln. The list for the Forth and Clyde canal comprises 20 locks with their fittings and features, 7 bascule bridge sites, 10 bridges of the 1960s period built using concrete pipes, 2 bridges with steel pipes, 5 concrete bridges and 5 steel bridges. A hand list of sites is given in the gazetteer below.

Gazetteer of sites recorded, A) on the Union Canal, and B) on the Forth and Clyde.

A)

1; Lemington Lift Bridge, Edinburgh.
Reinstated steel lift bridge with new lifting machinery.

2; Kingsknowe Pipes, Edinburgh.
Concrete pipe bridge.

3; Sheet piles with by-pass pipe at east end of blocked canal, Wester Hailes.
Details of sheet piles and by-pass pipes.

4; Dumbryden bridge, Wester Hailes.
Details of masonry bridge.

5; Wester Hailes Rd. Bridge, Wester Hailes.
Details of bridges from several periods.

6; Sheet piles with by-pass pipe at west end of blocked canal, Wester Hailes.
Details of sheet piles and by-pass pipes.

7; Tow Path, nr. Ratho.
Details of tow path surface.

8; M8, nr Broxburn.
Basic details of concrete pipes.

9; Stable, Bridge 23, Broxburn.
Details of stable built on to bridge abutment.

10; Greendykes Rd. Bridge, Broxburn.
Details of original masonry bridge and concrete pipes of the 1960s.

11; Bridges 28 and 29, Broxburn.
Details of masonry bridges.

12; Mile Stones, Winchburgh and Broxburn area.
Details of milestones still in position.

13; Bridge 39, Philpstoun.
Details of timber safety gate.

14; Vallore Rd. Bridge, nr. Linlithgow.
Details of masonry bridge with safety gate and later steel culvert.

15; Bridge 52, nr. Linlithgow.
Tow path details.

16; Lathallen Rd, A801, nr. Linlithgow.
Concrete pipe bridge details.

17; Lime kiln, Port Maxwell, Falkirk.
Details of lime kiln remains.

B)

1; Lock 3, Falkirk.
Details of lock remains.

2; Spillway and By-pass pipes by Lock 3, Falkirk.
Details of spillway.

3; Orchardhall Bridge, Falkirk.
Basic details, railway swing bridge.

4; Abbots Rd. Bridge, Falkirk.
Details of concrete pipe bridge.

5; Lock 4, Falkirk.
Details of lock gates and sluice gearing.

6; Lock 5, Falkirk.
Details of upper cill and lock gates.

7; Bainsford bridge, Falkirk.
Details of earlier bridges and concrete pipes.

8; Carron Iron Works Basin, Bainsford Bridge, Falkirk.
Details of former basin remains.

9; Lock 6, Falkirk.
Details of lock gates.

10; Lock 7, Falkirk.
Details of lock gates.

11; Lock 8, Falkirk.
Details of lock, upper cill, walls, lock gates and bridge.

12; Lock 9 and Railway Bridge, Falkirk.
Details of lock gates and railway bridge.

13; Lock 10, Falkirk.
Details of lock gates.

14; Lock 11, Falkirk.
Details of lock gates and upper cill.

15; Lock Keeper's Cottage, Lock 11, Falkirk.
Details of cottage remains.

16; Camelon Rd. Bridge, Falkirk.
Details of several bridges and concrete pipes.

17; Landing Stage and Basin Entrence, Glenfuir Rd. Falkirk.
Details of landing stage and basin entrence.

18; Lock 12, Falkirk.
Details of lock gates.

19; Lock 13, Falkirk.
Details of lock gates.

20; Lock 14, Falkirk.
Details of footbridge and lock gates.

21; Lock 15, Falkirk.
Details of lock gates.

22; Lock 16, Falkirk.
Details of lock gates, flanking walls road bridge.

23; Roadbridge, Bonnybridge.
Details of earlier bridge remains and later steel culvert.

24; Lock 19, Castlecary.
Detail of lock gate swing bar stop.

25; Castlecary Rd. Bridge, Castlecary.
Details of former bridges.

26; A80, Castlecary Arches.
Details of concrete pipes and underpass.

27; Wyndford Lock, Banknock.
Details of bridge at site of bascule bridge.

28; Auckinstarry Bridge, Auckinstarry.
Details of steel swing bridge.

29; Twechar Bridge, Twechar.
Details of steel lift bridge.

30; Hillhead Bridge, Kirkintillock.
Details of steel swing bridge.

31; Townhead Bridge, Kirkintillock.
Details of earlier bridge and concrete pipes.

32; Cadder Bridge, Bishopbriggs.
Details of modified bascule bridge.

33; Balmuildy Rd. Bridge, Bishopbriggs.
Details of two bridges

34; Lock 24, Glasgow.
Details of lock gate swing bar stop.

35; Clevden Rd. Bridge, Glasgow.
Details of bascule bridge and steel culvert.

36; Locks 28 to 30, Glasgow.
Details of masons's marks.

37; Sheet Piles, By-pass Pipes & Lock 31, Glasgow.
Details of sheet pile work, pipes and lock.

38; Lock 32, Blairdardie Rd, Glasgow.
Details of lock and canal.

39; Great Western Rd. Bridge, Glasgow.
Details of former lift bridge.

40; Spillway and By-Pass Pipes, Great Western Rd. Glasgow.
Details of spillway and pipes.

41; Sheet Piles, By-pass Pipe, and Lock 36, Glasgow.
Details of sheet piles, pipe and lock.

42; Duntreath Rd. Bridge, Underpass, Spillway & Pipes, Glasgow.
Detail of underpass, spillway and pipes.

43; Argyll Rd. Bridge, Clydebank.
Details of concrete road bridge.

44; Mall Foot Bridge, Shopping Centre, Clydebank.
Basic details of foot bridge.

45; Kilbowie Rd. Bridge, Clydebank.
Details of earlier bridge and concrete pipes.

46; Dalmuir Rd. Bridge, Clydebank.
Details of earlier bridges and concrete pipes.

47; Erskine Ferry Rd. Bridge, Clydebank.
Details of steel swing bridge.

48; Lock 38, Bowling.
Details of lock gate swing bar stop.

Hydraulic engineering chronology of navigable waterways

DR. H. C. PREUL
Emeritus Professor & Consultant, Civil & Environmental Engineering Dept.,
University of Cincinnati; Cincinnati, Ohio, USA

INTRODUCTION

Waterways and canals were certainly the cradle for developing the expertise of hydraulic engineering as a major technical arm in Civil Engineering. The terms canals and waterways are used here interchangeably. As background, a chronology listing of canal engineering achievements is given. Early hydraulic engineering concepts along with canal construction equipment and methods are outlined. The construction of canals created a need for a better scientific understanding of the principles of fluid mechanics and hydraulics. Therefore many of the scientific advancements in these fields followed canal development as can be seen in the theoretical relationships currently in use. As long practiced, the design of canals and waterways requires a thorough and special understanding of open channel flow hydraulics along with hydrology. In the 20th century, the basic technology developed in historic canal systems has advanced to vast river system waterways for barge transport as well as for recreational navigation. The Ohio River system in the USA is cited as an example. Principal engineering considerations in canal and waterway design are outlined. The paper concludes with a review of the evolution of hydraulic engineering in navigable waterways.

WATERWAY ENGINEERING HISTORY

Waterways have long been recognized as an important practical method for transporting commodities. As far back as 4$\pm$ millenniums B.C., history records the use of navigable waterways in Egypt and Babylonia in the form of barges towed by slaves in a water filled ditch. The ancient Grand Canal in China, one of the longest in the world, employed locks to raise and lower boats. Great hydraulic engineering feats have followed such as the Suez Canal, the Panama Canal, the Erie Canal in the USA, and many others. With the advent of railroads, canal use suffered a major decline. However, by applying the old canal system hydraulic concepts, major rivers have been fitted with locks and dams. Now waterway transportation has taken on new economic importance generally competitive with railways. Although many of the old canal systems have been abandoned, some countries, especially the U.K, have maintained major parts of their old systems and have seen a resurgence in canals especially for recreational use. An excellent example is the recent restoration of the Forth & Clyde and Union Canals in the U.K.

The Millennium Link. Thomas Telford, London, 2000, 38–46

General Chronology of Canal and Waterway Engineering

Following is a general chronology for viewing advancement in the hydraulic engineering of canals and waterways. Canals were the fore-runners of major river waterways. It is not intended as a complete listing.

Historic Canal and Waterway Achievements

4000 B.C.	Upper Egypt, King Menes Canal
2200 B.C.	Mesopotamia Shatt-el-hai Canal, linking Tigris and Euphrates Rivers
1700 B.C.	Upper Egypt, Wadi Haifa Canal
600 B.C.	China, Wild Goose Canal, 419 km (260 mi) linking Yellow and Huai River
103 B.C.	China, Romans, Arles, France Canal to Mediterranean
66 A.D.	Nero attempts/abandons Corinth Canal through Greece's Isthmus
120	Britain, Romans build the Foss Dyke
600-610	China, construction of Grand Canal
983	China, invention of chambered lock on Grand Canal
1065	The Dutch build canals with single locks
1373	The Dutch build a primitive double lock on River Lek at Vreeswijk
1566	England, Exeter Ship Canal completed
1500 - 1800s	Canal construction flourishes in the U.K., Europe and other countries
1788	England, inclined plane first used to raise canal boats, Ketley Canal
1859	Suez Canal Completed
1888	France: Sensee Canal at Fontinettes, vertical boat lift replaces 5 locks
1893	Greece, Corinth Canal finally completed
1956	Egypt nationalizes Suez Canal
1903-14	USA: Construction of Panama Canal
1959	USA-Canada: St Lawrence Seaway opens to traffic
1993-2009	China: constructing the great Three Gorges Dam with 5 shiplocks for 10,000 ton navigation vessels and one shiplift for 3000 ton vessels

Canal Chronology (abbreviated): United Kingdom

1500s-1800s	Canal construction flourishes
1768-1790	Construction of Forth & Clyde Canal; world's first man-made sea to sea canal
1800s-1900s	Railways begin decline late 1800s; restoration and preservation take hold
2000	Edinburgh, Scotland, Millennium Link Conference signals canal resurgence

Canal and Waterway Chronology (abbreviated): USA

1716-1730	First canals; New York State Canal built across neck in Mohawk River
1730-1870	Canal development flourishes
1817-1825	Construction of Erie Canal
1870-71	Pennsylvania Railroads buy out canal navigation companies …… signaling decline of canals in USA
1903-14	Construction of Panama Canal
1910-29	46 locks & dams completed in canalization of Ohio River
1930	28 locks & dams authorized along upper Mississippi River
1959	USA-Canada: St Lawrence Seaway opens to traffic
1900-2000	Canals replaced by railroads; preservation and restoration old Canal Systems
1910-2000	Emphasis on development of major river waterways
1972-1984	Tennessee-Tombigbee Waterway constructed

EARLY HYDRAULIC ENGINEERING OF CANALS

Early canal building is seen to have evolved with the hand digging of diversion ditches from rivers such as the Nile in Egypt and the Euphrates and Tigris in Mesopotamia. These ditches likely were initially for irrigation and water supply purposes, then became an obvious method for floating materials from one location to another. Soon this extension further developed into the construction of canals for commercial transport, navigation, sailing, irrigation, water supply, washing clothes, bathing, and related purposes. The earliest surviving map of canals is inscribed with a reed stylus on a clay tablet, showing the two great canals flowing through the ancient city of Nippur which stood at the great fork of the Euphrates. In these arid countries, canals became of great importance in the wealth of Empires and Kingdoms. Construction of these early canals was the beginning of hydraulic engineering. The planning of a canal project required some understanding of fluid mechanics including concepts of fluid statics and flow of water in a channel. Plans for significant projects certainly were drawn or sketched in some form by an "hydraulic artisan" of recognized skill who then worked with other civil artisans in constructing the project. As commissioned by Pharaohs and Kings, these were the fore-runners of "turn-key" projects and "design-builders". It is expected that early "hydraulic artisans" understood and employed the following concepts in their professional practice:

Early Known Practical Hydraulic Engineering Concepts

- Hydraulic gradient: water flows down hill
- Water storage volume = f(L x W x H)
- Static pressure in a tank or body of water = f(depth)
- Flow of water in a channel = f(Area, shape, bottom slope, smoothness)
- System of quantifying Units: foot length, hand width, etc.
- Siphon principle

Early Engineering Equipment:

- A tablet of stone or papyrus for inscribing a plan
- A marked off standard rule or rod for laying out dimensions
- A rope or chain for laying out extended distances
- A plumbing rod of some type for establishing vertical alignment
- A leveling device in the form of liquid contained in a vessel for establishing slope or deviation from the horizontal
- On site siting table for establishing alignment

Early Construction Equipment and Methods

- Hand shovels and digging
- Scoops or slips pulled by teams of oxen or men
- Inclined planes for elevating or lowering materials
- Ropes and lines
- Tripods for lifting and lowering materials
- Pulleys for lifting and lowering materials

GREEK AND ROMAN CANAL BUILDERS

Although the Egyptians and Babylonians were practical fore-runners, the Greeks and Romans significantly advanced scientific concepts of fluid behavior into basic written relationships. Numerous methods of handling fluids were invented, i.e., Archimedes' screw, water clock, pumps, siphons, etc. Fluid concepts were quantified so that projects could be scientifically planned and constructed. The flow of a water volume was related to time, **Q = Vol/time.** Such basic understandings of fluid mechanics became the theoretical foundation for the application of hydraulic engineering, such as in the design of the Corinth canal which was an original dream of Emperors, including Nero, but only finally completed in the late 1800s. The Romans in particular built vast water supply aqueducts and developed canal systems. Leonardo da Vinci was truly an hydraulic engineer in his understanding and planning of river control schemes and construction of canals. Largely his contribution was as a conceptual designer and inventor. He recorded plans for a number of large river and canal works, some of which were constructed. Leonardo advanced the concept of continuity in river flow as follows : "A river in each part of its length in an equal time gives passage to an equal quantity of water, whatever the width, the depth, the slope, the roughness, the tortuosity. Each movement of water of equal surface width will run the swifter the smaller the depth.... And this motion will be of this quality: .... A river of uniform depth will have a more rapid flow at the narrower section than at the wider, to the extent that the greater width surpasses the lesser." Leonardo well described continuity as we now use it:

$\mathbf{Q = (AV)_1 = (AV)_2}$ as applied to waterways and canals

Note: Notation definitions are given at end of paper.

ADVANCEMENTS IN CANAL AND WATERWAY HYDRAULICS: 18TH CENTURY

During this period, experimental hydraulics advanced concepts of fluid mechanics and hydrodynamics. Navigable Canal construction flourished around the world for commercial transport. Great systems were constructed in England, Europe, the United States, and many other countries. With advancements in locking systems, canal pools became longer and accommodated higher lifts.

Theoretical Equations which Advanced the Hydraulic Engineering of Canals:

Poleni (1738): equation for sharp crested weirs:

$$Q = (2/3)\, C\, b\, (2g)^{1/2}\, h^{3/2}$$

Bernouilli energy principle (1738): (open channel flow form):

$$Y_1 + (V^2/2g)_1 + Z_1 = Y_2 + (V^2/2g)_2 + Z_2 + h_L$$

Chezy (1768): equation for calculation of open channel and canal discharge: final form:

$$V = C(RS)^{1/2}$$

Lagrange (1788): Motion of a solitary wave of initesimal height in a channel of finite depth:

Velocity of propagation: $v = (gy)^{1/2}$

Woltman (1790): application of spoke-vane type current meter

ADVANCEMENTS IN CANAL AND WATERWAY HYDRAULICS: 19TH CENTURY

Based on experimental data, a number of empirical relationships were produced applicable to open channel flow:

Eytelwein (1801): open channel flow formula (in metric units):

$$V = 50.9\, (RS)^{1/2}$$

Bidone (1820, 1826): discovered "jump of Bidone" known as "hydraulic jump"

Belanger (1849): Discharge over a broad-crested weir:

Russell (University of Edinburgh)(1842-43): Conclusions on the importance of wave formation in the resistance of ships. In 1834, he undertook for a local canal company an investigation of the practicality of steam navigation on inland waterways.
Dupuit (1848): published theories and descriptions of surface profiles for open channel flow:
Bazin (1865): described variation in velocity distribution with relative depth of flow
Manning: (1889): For flow of water in open channels and pipes:

$$\mathbf{V = (1.49/n)\, R^{2/3}\, S^{1/2}}$$ as applicable for English units

Moritz Weber (1919): applied Froude Number for defining types of open channel flow with reference to critical flow (Sub-critical, Super-critical): $\mathbf{N_F = V/(gy)^{1/2}}$

ADVANCEMENTS IN WATERWAY HYDRAULICS: 20TH CENTURY: USA

In the latter part of the 19th century, railways largely replaced canal systems in many countries, especially in the USA where the smaller tributary canals fell into disuse except for limited reaches and locks kept as historical parks and museums. Some larger canal systems have been maintained for recreational boating and commercial transport. The State of New York still maintains a canal system including the famous Erie Canal which runs from the Hudson River to Lake Erie. After the turn of the 20th century, USA inland waterway transportation mainly turned toward development in major rivers and harbors. These waterways became the new canal systems. In the USA, under the direction of the USA Army Corps of Engineers, there are now around 40,250 km (25,000 mi) of navigable waterways. Although these river waterway systems now have been constructed to accommodate large barge tows carrying great tonnage, they still emulate the original operations of the old canal systems.

Advancements in Waterway Hydraulics:

- Open channel flow concepts have been refined for a wide range of applications
- Unsteady flow theory and practice have been significantly advanced, i.e., flood routing
- Computerized methods have greatly advanced the analysis of open channel flow systems
- Although computer simulation techniques have been a great advancement with speed of computation, laboratory experimentation has suffered greatly since about year 1970.

HYDRAULIC ENGINEERING OF CANALS AND WATERWAYS: USA

Hydraulic engineering for the design and construction of canals in the USA started with experienced canal engineers who came from England and the Netherlands. These European engineers trained the first generation of American professionals in canal building from the late 1790s mostly in the New York area. From this small Corps of Engineers grew the technology for the beginning of the Erie Canal in 1817 and the training of younger associates who moved on to Pennsylvania, Ohio, and other states. The design of canals and waterways requires special training in a broad range of hydraulic concepts as listed above along with a thorough knowledge of hydrology.

Basic Engineering Considerations in Design of Canals and Waterways

- Topography and route between water bodies
- Elevation differences
- Water supply and regulation
- Hydrology including flood control
- Vessel size
- Required channel size and depth
- Channel slope and roughness
- Channel erosion and siltation problems
- Channel seepage and losses
- Number and size of locks
- Consumer demand and competition
- Operating economy

Figure 1:

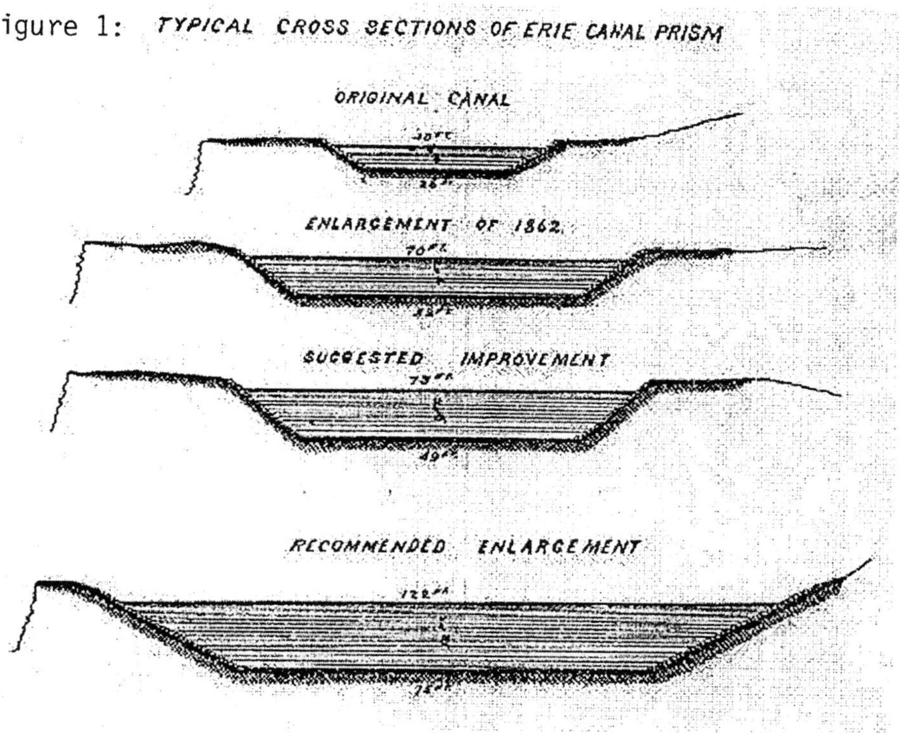

Figure 1 shows typical prismatic channel cross sections of the original Erie Canal and enlargements. Trapezoidal side slopes varied according to composition of soil materials.

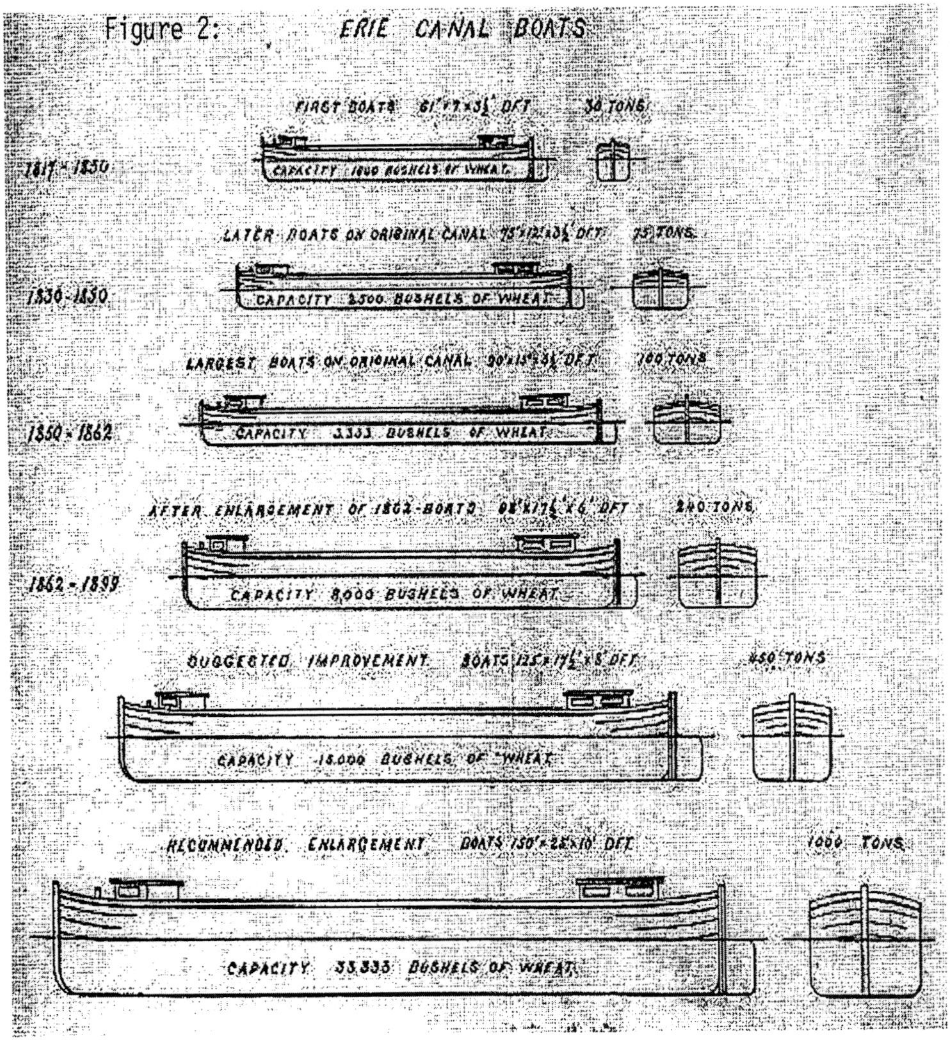

Figure 2 shows a series of Erie Canal Boats from 1817 to 1899.

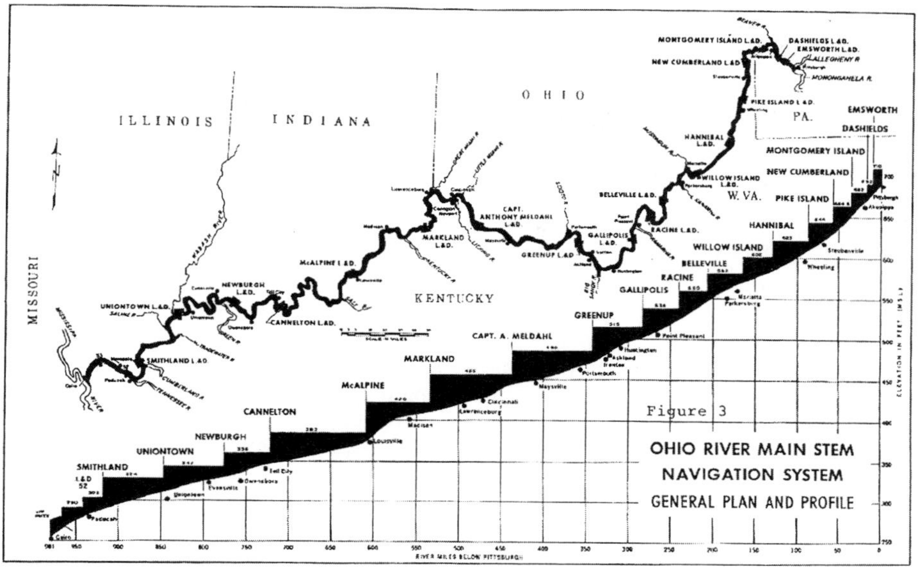

Figure 3 shows a profile of the Ohio River Main Stem Navigation system including 20 locks and dams with locks lifts varying from approximately 3 m (10 ft) up to 10.7 m (35 ft). A 3.4 m (11 ft) channel depth is maintained for barge traffic. Locks accommodate both commercial barge traffic and recreational boating.

BACK TO THE FUTURE: WATERWAY ENGINEERING - 21ST CENTURY

The Millennium Link signals an innovative future in waterway and canal engineering. In looking back and now to the future, navigable waterways are viewed here in two general categories: waterways for purposes of commercial transport combined with recreational use; and waterways/restored canals for recreation, pleasure boating and as historical monuments. Commercial waterways must compete for use and development based on prevailing economics, even in the 21ST century. Recreational waterways and restored canals with their special nostalgia are expected to gain significantly in popularity. Restorations will continue to require specialized engineering and construction expertise. In general, waterways will greatly benefit from technology advancements envisioned as outlined below:

- Improved communication including extensive application of global positioning technology for vessel and facility operations
- Increased size and efficiency in locks on major commercial waterways as older facilities require replacement
- Automatic elevator systems and innovative wheels replacing locks for faster transport, especially for smaller navigation vessels
- Innovative dredging machines with high tech guidance systems for channel maintenance
- Increased use of hydraulic turbines for water power especially on major river systems

- Increased recreational use: i.e., competitive rowing, wind sailing, skiing, pleasure boating
- Extension of fisheries with improved water quality
- Historical parks and recreational trails especially along restored canal routes

As an example for the future, the U.S. Corps of Engineers forecasts waterway traffic on the Ohio River system to have an annual growth rate of about 1.1 percent through year 2060, a total of over 60%. Improvements in locks to more efficiently accommodate navigation are a further expectation.

SUMMARY AND CONCLUSIONS

Hydraulic engineering was certainly one of the first civil engineering specialties beginning with early canal building. The design and construction of canals created a need for a better understanding of the flow of water and consequently was a great driving force in the development of current concepts of fluid mechanics and hydraulics. These concepts are in the form of theoretical and empirical relationships to be found in University level textbooks and practical engineering references. Canals have long been of great economic importance throughout the world withstanding flourishing periods and a period of decline with the advent of railroads in the 1800s. Following their decline, especially in the USA, the old concept of canals changed to great waterways concentrated in major rivers especially for commercial transport as well as for pleasure craft. Concurrently with the development of major river waterways, there has been a resurgence in the preservation of parts of the old canal systems in some cases for continued economic service and in others as historical museums.

In the U. K., the rehabilitation and reopening of the Forth & Clyde and Union Canals along with the Millennium Wheel signal a new appreciation in the hydraulic engineering of canals and waterways.

NOTATION

f	function of listed factors
Q	channel flow
A	cross sectional area of channel
V	average velocity
C	empirical constant
b	breadth of broad crested weir
g	gravitational constant
h	head
Y or y	channel depth
Z	elevation reference, potential energy
$\mathbf{h_L}$	head or energy losses
R	hydraulic radius
S	energy gradient or channel slope
n	channel roughness
$\mathbf{N_F}$	Froude number

ACKNOWLEDGMENTS

It is gratefully acknowledged that a number of important references and sources have been utilized in the preparation of this paper, mainly as follows:

(1) Minor, David. 1996. Eagles Byte Research; http://home.eznet.net/~dminor/Canals.html.

(2) Rouse, Hunter and Ince, Simon. 1957. History of Hydraulics. Iowa Institute of Research, State University of Iowa.

(3) New York State Canal System. Internet http://www.history,rochester.edu/canal/

(4) U.S. Corps of Engineers, Ohio River Division, Cincinnati, Ohio.

American canals as a source of revitalization: the Chesapeake and Ohio Canal as a case study

ROBERT J. KAPSCH, Ph.D., ASCE, Hon. AIA
National Park Service, Washington, D.C.; U.S.A.

INTRODUCTION

The Chesapeake and Ohio Canal (C & O) constructed between 1828 and 1850, spans much of the canal era in the United States. Begun after the Erie Canal was opened, it continued in commercial operation until 1924. The C & O Ohio Canal never reached the Ohio River but terminated in Cumberland, Maryland, approximately 296 km (184 miles) northwest of Washington, D.C., the beginning point of the C & O. The C & O was conceived to be the grandest of the long distance canals then built. It was to replace the older Patowmack Canal Company, begun by George Washington, a series of small bypass canals and slack water navigation. The C & O, in contrast, was planned to provide stillwater navigation from the Chesapeake Bay to the Ohio River. Both the C & O and its predecessor, the Patowmack Canal, shared George Washington's vision of utilizing the Potomac River as a corridor to the wealth of the trans-Appalachian Region.

On the day when the C & O Canal was begun, July 4, 1828, approximately 80 km (fifty miles) north, in Baltimore, Maryland, another group of men began the country's first long distance railroad – the Baltimore and Ohio Railroad (B & O). The early construction history of the C & O Canal is a story of intense legal and physical competition with the B & O Railroad for access to the narrow mountain passes in western Maryland where both canal and railroad had to pass. It is also a story about the end of the canal era and the beginning of the railroad era in America.

The first Chief Engineer for the C & O Canal was Benjamin Wright, the Chief Engineer of the Erie Canal. Wright is known as, "The Father of American Civil Engineering" because American canals in general and the Erie Canal in particular became the training ground for talented men in this new profession. When Wright became Chief Engineer of the C & O Canal, he brought many of his assistants with him from the Erie. Although only Chief Engineer of the C & O Canal for a short time, from 1828 to 1830, Wright left his mark on the construction of the canal.

At the beginning of the construction of the C & O Canal, it was not apparent that railroads were to eclipse canals for long distance haulage. The promoters of the C & O Canal believed that a superior designed and constructed canal, along with steam power, was to be the future of long distance haulage in America.

The Millennium Link. Thomas Telford, London, 2000, 47–54

In 1938 the C & O Canal came into Federal ownership, purchased from the then owner, the B & O Railroad, utilizing Public Works Administration funds as part of the Federal Government's depression-relief effort. The National Park Service was assigned the responsibility of caring for this new acquisition. World War II prohibited immediate work but after the war the National Park Service proposed converting the canal into a Parkway. The public, in 1954, led by Chief Justice William O. Douglas, protested and the plans were abandoned. These efforts led to additional efforts for creating a National Park of the C & O Canal and, on January 8, 1971, President Nixon signed legislation establishing the Chesapeake and Ohio Canal National Historical Park.

Since its establishment the Chesapeake and Ohio Canal National Historical Park has been a success story. The Park protects the country's early industrial history, provides a natural refuge for numerous endangered species, and provides recreation opportunities for hiking, biking, horse back riding and numerous other activities. It has also been a source of revitalization for the towns through which it passes. Perhaps most importantly, as the country's oldest canal-turned-park, it provides a model for the dozens of other canals throughout the nation that have been turned into parks, recreation opportunities and sources for revitalization.

BENJAMIN WRIGHT AND THE BEGINNING OF AMERICAN CIVIL ENGINEERING[1]

Benjamin Wright was born October 10, 1770 in Wethersfield, Connecticut. His early career was primarily that of a surveyor in New York State. In 1790, although Wright did not work for him, he was undoubtedly aware of the work of the English Engineer William Weston (1753-1833) and his work in making canal surveys for the Western Inland Lock & Navigation Company near Rome, New York and along routes that were to be later used by the Erie Canal. Later, Wright would work for the same company on the work at Little Falls on the Mohawk River (built 1793-95) and in the construction of the six-mile Wood Creek canal (1803). He was engaged in number of other surveying and navigation projects, primarily centered on the Hudson River.

Wright never visited the canals of the United Kingdom but was aware of them through his study of the 80-page canal entry in Rees' Cyclopedia. He had visited the Middlesex Canal near Boston and was aware of that construction. Following the conclusion of the War of 1812 with England, canal construction greatly increased. On July 4, 1817 ground was broken for the Erie Canal. The commissioners invited William Weston to become Chief Engineer of the work but he declined. Reluctantly they offered the position to Wright.

The Erie Canal became a training ground, under Wright, for American civil engineers – many, such as James Geddes (1763-1838) and Nathan Roberts (1776-1852) were to go on to work under Wright on the C & O Canal. The construction of the Erie Canal provided the impetus for the first discovery of hydraulic cement in America, discovered by Canvass White. Although Wright never made a trip to England, White did, in 1818, and made many

---

[1] This portion of the paper draws heavily on Neal FitzSimons' work on Wright, "Benjamin Wright: The Father of American Civil Engineering," published in Jerry R. Rogers, et.al., Civil Engineering History: Engineers Make History, Proceedings of the First National Symposium on Civil Engineering History, November 10-14, 1996, (New York: American Society of Civil Engineers, 1996), pp. 99-107. Mr. FitzSimons was scheduled to address this conference but unfortunately died, on March 23, 2000. This paper is dedicated to his memory and in recognition of his many contributions to civil engineering history.

detailed drawings of the English canals. On his return to America White discovered hydraulic cement near Chittenago, New York. The experience gained with hydraulic cement on the Erie Canal would later be used by Wright in developing hydraulic cement at Shepherdstown, Virginia (now West Virginia) and at Tuscarora Creek, Maryland, for support of the construction of the C & O Canal.

The Erie Canal was opened on November 7, 1825, and Wright turned to other tasks, such as work on the Delaware and Hudson Canal, the Chesapeake and Delaware Ship Canal, the James and Kanawha and the New York Waterworks Company and other projects. In 1828 he began work on the Chesapeake and Ohio Canal as its Chief Engineer, at age 58.

BENJAMIN WRIGHT AND THE CONCEPTUALIZATION OF THE CHESAPEAKE AND OHIO CANAL

On September 4, 1828 Benjamin Wright was offered the position of Chief Engineer of the C& O Canal by the Company President Charles Fenton Mercer at an annual salary of $5,000 per year.[2] In fact, in an informal arrangement with Mercer, Wright had been working, along with his assistant, Nathan Roberts, for many months. Working seven days a week, by July 1828, Wright and Roberts had developed the surveys, contracts and specifications for the first 27 km (17 miles), laid out in 0.8 km (half mile) sections, of the canal, from Little Falls to Great Falls, Maryland. These contracts were awarded by August 1828. The C & O Canal Company received 462 proposals and awarded $218,000 worth of contracts in that month, most to experienced canal contractors from New York and Pennsylvania.[3] By October 1828, Wright and his assistants had an additional 47 km (29 miles) under contract, from Great Falls to Point of Rocks, Maryland. An additional portion, from Georgetown in Washington, D.C. to Little Falls, consisting of eight sections and four lift locks encompassing 8 km (5 miles) was also put under contract This work included not only the excavation (and fill) for approximately 80 km (50 miles) of canal, but also the construction of twenty-seven (27) masonry 2.8 m. (8 foot) lift locks, two substantial masonry aqueducts (Seneca and Monocacy) and numerous culverts. Within a year approximately 6,000 men and 700 horses would be working on the line of the C & O Canal extending from Georgetown in Washington, D.C. to Point of Rocks, Maryland.[4]

To administer this work, Wright established a Board of Engineers organized into three divisions – the first Division encompassing the work put under contract in 1828 and the remaining two Divisions encompassing the remainder of the C & O Canal, from the Blue Ridge Mountains to Pittsburgh, Pennsylvania. The First Division was organized into five residencies each with a principal engineer, an assistant engineer and a rodman.[5] The principal function of the engineers of the First Division was to oversee and certify work undertaken by the various contractors.

As ambitious as this sounds, there is ample evidence that Wright intended to undertake twice this level of effort -- he intended to have approximately 160 km (100 miles) under contract in the first year rather than 80 km (50 miles).[6] What stopped him was the injunction of the

---

[2] Records of the Chesapeake and Ohio Canal Company, Letters Sent, Record Group 79, National Archives.
[3] Ibid., First Report of the President and Directors to the Stockholders of the Chesapeake and Ohio Canal Company, Proceedings of the President and Directors, September 10, 1828.
[4] Ibid., Third Annual Report, June 6, 1831.
[5] Ibid., Proceedings of the President and Directors, November 22, 1828.
[6] Ibid., Letters Received, Wright to Ingle, October 31, 1828.

Chancellor of the State of Maryland, in response to a law suit filed by the Baltimore and Ohio Railroad, enjoining and prohibiting the C & O Canal Co. in "obstructing or hindering the Baltimore and Rail Road Company in the use of the rights obtained by them…as to route."[7] In effect, this injunction prohibited the C & O Canal Co. from initiating construction beyond Point of Rocks on the Potomac River. Point of Rocks was a narrow passage along the east side of the Potomac River, approximately 80 km (50 miles) from Washington, D.C. and also approximately 80 km (50 miles) from Baltimore. If the C & O Canal Co. could not use this narrow passage at Point of Rocks than an expensive river crossing of the Potomac River would have to be made. Likewise for the Baltimore and Ohio Railroad. The legal battle between the two companies continued for almost five years and was not resolved until January 1832, when the canal company won the suit on appeal.[8] In the meantime, construction was only begun by the canal company on that portion of the canal between Washington, D.C. and Point of Rocks, Maryland. Until the legal issue between the canal and railroad was resolved, the narrow passage at Point of Rocks was the cork in the bottle to C & O Canal construction operations.

In constructing the first fifty miles of the C & O Canal, Wright was encouraged by C & O Canal Company President Mercer to build an exemplary canal such as had not been previously seen in the United States. Mercer and Wright agreed upon a canal whose prism dimensions would exceed long distance canals then built in the U.S. – at its surface the canal's prism was to be 18.3 m. (60 feet) wide; it was to be 1.8 m. (6 feet) deep and to be 13.7 m. (45 feet) wide at its bottom. And the line of canal would avoid all abrupt curvatures. The bank linings, where practical, were to be paved with stone to save future repair.[9] What Mercer and Wright were thinking about was steam-propelled vessels on the C & O Canal. In 1830 the Board of Directors approved an order to advertise for, "a design for a boat suited to the navigation of the Canal both and with-out steam power; and also for proposals for constructing the same, including a suitable steam-engine."[10] A month later they authorized Mercer to, "request of Commodore John Rodgers a plan for the structure of a Canal boat."[11] And in their third annual report, of 1831, they reported on, "the interesting experiments on the canals of Scotland, where a velocity of from 10 to 15 miles per hour has been attained without any surge injurious to the (canal) banks … and which (therefore) place canals, in point of attainable speed, on an equal footing with rail-roads."[12]

Throughout the history of the C & O Canal, canal barges traveled at the speed of mules and horses and were not steam powered. Nonetheless Wright and Mercer were thinking in terms of a canal system that would expedite cargoes faster than normal canals. They also looked at the time that it would take a canal boat to transit a lift lock as a means of speeding cargoes. There were seventy-five (75) lift locks to be built on the C & O Canal between Washington, D.C. and Cumberland, Maryland – mostly of 2.4 m. (8 foot) lift. The locks were of very generous dimensions by contemporaneous standards – the chambers of these locks measuring 30.5 m. x 4.6 m. (100 feet by 15 feet). The canal company undertook experiments to reduce the transit time. These experiments indicated that a boat could transit these locks in 2 – ½

---

[7] Niles' Register (Baltimore), June 28, 1828.
[8] Records of the Chesapeake and Ohio Canal Company, Proceedings of the President and Directors, January 7, 1832, Record Group 79, National Archives.
[9] Ibid., September 10, 1828.
[10] Ibid., February 5, 1830.
[11] Ibid., March 13, 1830.
[12] Niles' Register, Baltimore, July 16, 1831.

minutes each.[13] Wright believed that the time of transit could be further reduced to 2 minutes.[14]

To further encourage speed and to facilitate the possible use of steam boats on the C & O Canal, Mercer adopted a policy of no bridges across the canal, except in Georgetown where it would be impossible to avoid. The C & O Canal company, primarily through Wright's actions, went to great lengths to avoid building bridges. Where possible, the company bought land between the canal and the river so as to eliminate the need for farmers to cross the canal. Where this was not possible, the company built pivot bridges and, where necessary, (such as the access road to Chain Bridge just northwest of Washington, D.C.), they built large road culverts under the canal.[15] These road culverts are still used for pedestrian and, in some cases, vehicular traffic. At ferry crossings of the Potomac River, Mercer issued instructions to the engineers to build ramps in the canal bed so that during the winter when the canal was drained wagons could easily cross the dry canal. For operating season, Mercer ordered the engineers to leave enough room to construct a basin for a scow large enough for a wagon and a team of six horses to ferry across the canal.[16]

The C & O Canal was designed by Wright to greatly increase its boat carrying ability. Wright, for example, issued orders to his engineers to site all of the locks to one side of the line of the canal so that, when necessitated by the expected increase in canal traffic, an additional lock could be constructed adjacent to the first lock thereby allowing canal boats to both go up and down the canal at the same time and without waiting.[17] In the heady first year of construction, the Board of Directors issued an order to the Board of Engineers that an estimate be produced to double the size of the Monocacy Aqueduct, the largest on the line of the canal at 157.3 m. (516 feet) in length, "...so as to admit the free passage of boats moving upon it, at the same time in opposite directions."[18] The cost estimate developed by the engineers is not in the surviving records, but it must have been a very large one. The Monocacy Aqueduct was constructed only wide enough for a single boat. Even then it took the C & O Canal Co. almost five years to construct it.

Mercer and Wright were in agreement to build the C & O Canal in the finest manner. The first 80 km (50 miles) of the C & O Canal included two aqueducts, Seneca and the Monocacy. The company considered less expensive alternatives to the masonry arch structures designed by Wright, such as wooden aqueducts and suspension bridges, but never adopted these. Unlike the Erie Canal, the C & O Canal never used dams to impound water for crossing streams, in lieu of aqueducts. The insistence of quality masonry construction, by Mercer and Wright, can be seen today, some 160 years or so since these structures have been built.

---

13 Ibid., November 27, 1830.

14 Records of the Chesapeake and Ohio Canal Company, Letters Sent, Wright to William Archer, September 1, 1829, Record Group 79, National Archives.

15 The road culvert to Chain Bridge was 15.9 m. (52 feet) long and had a span of 3.7 m. (12 feet). It cost approximately $2,000 to build. Ibid., Letters Received, Wright to Mercer, October 3, 1829.

16 Ibid., Letters Sent, Mercer to M.C. Fairfax, December 22, 1828.

17 Ibid., Letters Received, Wright to W.M. Fairfax, December 8, 1828.

18 Ibid., Proceedings of the President and Directors, December 3, 1828.

BENJAMIN WRIGHT AND THE EARLY CONSTRUCTION OF THE C & O CANAL

The C & O Canal was being built through territory that was sparsely settled. Wright had three basic problems in the construction of the canal: securing an adequate supply of cut stone where quarries were not existent or inadequate, securing an adequate supply of hydraulic cement and in securing an adequate supply of labor, particularly skilled labor, such as masons.

Wright met the problem of an adequate supply of cut stone in a number of ways. For those locks in and near Washington, D.C., and therefore accessible by boat, Wright used Aquia stone from the Aquia quarries some 64 km (40 miles) south of Washington, D.C. These quarries dated from the seventeenth century and were used by the builders of the White House and the Capitol during the 1793 to 1800 Federal building construction program. Wright examined the stone on these buildings and found that the deterioration that was apparent was the result of fire set to these building by the British during the War of 1812 and not due to any inherent weakness of the stone itself.[19]

Further up the line of the construction of the C & O Canal, Wright used the stone from two existing quarries, the Granite Quarry at Little Falls and the red sandstone quarry at Seneca. But the greatest challenge was to find a source of stone for the Monocacy Aqueduct at the upper reaches of the 80 km (50 mile) construction effort. Wright calculated that the Monocacy Aqueduct would require 6000 cu. m. (8500 perches) of masonry at a cost of $57,000 – a huge cost by the standards of the day. A source of cut stone was found and a quarry was opened on the farm of Mrs. Nelson, approximately 6 km (4 miles) from the construction site. However the stone was inferior and Wright had to order tearing down the first three piers of the aqueduct. A quality white stone was found on the farm of Joseph Johnson not too far distant from the Monocacy Aqueduct and that stone was used in its construction.[20]

For hydraulic cement, Wright first used Parker's Roman Cement, shipped from England. But this was expensive. He augmented this with New York hydraulic cement, available at about half the price. But the problem of supplying adequate amounts of hydraulic cement to the contractors was not settled until the Shepherdstown, Virginia (now West Virginia) mill and bank of kilns was developed and hydraulic cement was produced there at less than one third the cost of the New York hydraulic cement and one fifth of the imported cement. This was later augmented by hydraulic cement produced on Tuscarora Creek, not far from the Monocacy Aqueduct.

The problem of skilled labor was resolved through the use of indentured servants, primarily from the British Islands. However, many of these left the C & O Canal to work for its competitor, the B & O Railroad. Cholera took a large toll on the workforce as well.

Wright's contributions to the C & O Canal not only included the overall engineering direction of the first fifty miles but also the design of individual structures such as the Monocacy Aqueduct. He also designed Section 15 (called, "The Great Embankment") and Section 18 (sometimes called, probably erroneously, Mary's Wall) at Great Falls which allowed the

---

[19] Ibid., Letters Received, Wright to Mercer, November 1, 1829.

[20] Robert J. Kapsch, "Benjamin Wright and the Design and Construction of the Monocacy Aqueduct," Canal History and Technology Proceedings, Lance Metz (Ed.), Volume XIX (Easton, Pa.: Canal History and Technology Press, 2000), pp. 181-222.

canal company to use the old bed of the Potomac River and to avoid extensive rock excavation. These structures were large: Wright estimated Section 15 as containing, "... about 60,000 Cubic Yard (46,000 cu m.) ... all in a short space of distance." [21] These structures have been called the largest dry laid stone retaining walls in America.

Benjamin Wright left the employ of the C & O Canal in October 1830 because, "... of the difficulties still subsisting between the Ches & Ohio Canal and the Balt & Ohio rail road ... there (being) very little field for active operations for Engineers left."[22] In the short time that he had been employed by the C & O Canal he had accomplished much. He kept in contact with Mercer and the Board of Directors over the next several years but his engineering practice turned from canals to railroads.

In April 1833, the Monocacy Aqueduct, the last unfinished structure on the 80 km (50 miles) of the C & O Canal was finally completed. That portion of the canal was watered later that year. Shortly after the completion of the Monocacy Aqueduct, C & O Canal President Mercer was voted out of office by the stockholders and replaced by John H. Eaton.[23] The stockholders then voted Mercer a payment of $5,000 to express their thanks[24] -- an early example of what has come to be called the golden handshake.

After the departure of Wright and Mercer, the C & O Canal continued its struggle to reach Cumberland, Maryland, which it did some seventeen (17) years later, in 1850. No attempt was made to continue over the mountains to the Ohio River. The C & O Canal continued in existence longer than most long distance American canals. But in 1924 halted operation after the most recent of a series of floods that ravaged the canal since its inception. Ownership of the C & O Canal reverted to its long time adversary, the B & O Railroad. In 1938 the Federal Government purchased the defunct C & O Canal from the B & O.

The C & O Canal as a Source of Revitalization

On one level the Chesapeake and Ohio Canal was unsuccessful. It never reached the Ohio River, its intended goal, and despite being long lived it was never a commercial success. But on another level, that of being a national park to meet the needs of citizens of the twentieth and twenty-first century, it proved to be very successful. Unlike many National Parks which are established for a single or a cluster of related values, such as scenic splendor or historical significance, the Chesapeake and Ohio Canal contains multiple values, such as natural resource protection, recreation and cultural resource interpretation. As a protector of natural resources, the Chesapeake and Ohio Canal contains numerous rare and endangered species, such as 160 different rare plants that exist within the Park. As a source of recreation, the Chesapeake and Ohio Canal provides 296 km (184 miles) of hiking, bicycle riding, horseback riding located in a spectacular natural river basin environment. It is linked to the nation's system of long distance hiking trails, the most famous being the Appalachian Trail. As an interpreter of cultural resources, it contains the most complete example of the canal era left in the United States. As the urban areas surrounding this Park become more developed, the value of the Park to its citizens becomes more valuable.

---

[21] Records of the Chesapeake and Ohio Canal Co., Letters Received, October 27, 1829, Record Group 79, National Archives.

[22] Ibid., Letters Received, July 31, 1830.

[23] Niles' Register, Baltimore, June 8, 1833.

[24] Records of the Chesapeake and Ohio Canal Company, Proceedings of the President and Directors, June 14, 1833, Record Group 79, National Archives.

The Chesapeake and Ohio Canal National Historical Park also serves as a focus of revitalization for the communities which it passes through, such as Georgetown, Brunswick, Williamsport, Hancock, Cumberland and others. The Canal provides a focus of attention and pride for these communities, each of which sponsor celebrations centered on the Canal and celebration of the Canal.

In one sense the Chesapeake and Ohio Canal was not a success. It lost to the B & O Railroad, never made it to the Ohio River and never proved as highly profitable as it backers supposed. On the other hand, the Chesapeake and Ohio Canal has proved to be a great success for millions of Park visitors who utilize it to commune with nature, watch birds, jog, bike, horseback ride, or contemplate America's industrial heritage. It has also proved to be a success in assisting communities along its line in revitalization.

Robert J. Kapsch, Ph.D., ASCE, Hon. AIA
National Park Service, Washington, D.C.: U.S.A.

Robert J. Kapsch is currently serving as Special Assistant to the Deputy Director, National Park Service. He undertakes special projects for the National Park Service, including special projects related to the Chesapeake and Ohio Canal National Historical Park. Prior to this assignment Dr. Kapsch was Chief of the Historic American Buildings Survey/Historic American Engineering Record (HABS/HAER) for fifteen years – the U.S. Government's principal historic documentation program for architecture and engineering. Dr. Kapsch has a bachelor's degree in civil engineering; master's degrees in management and historic preservation and doctorate degrees in architecture and engineering and American Studies.

The Millennium Link: economic impact

PETER WOOD, DTZ Pieda Consulting, Planning, Economic and Development Consultants, 28 Drumsheugh Gardens, Edinburgh, EH3 7RN

1. INTRODUCTION

THE PROJECT

1.1 The project consists of the restoration to operation of the Forth and Clyde and Union Canals to form the "Millennium Link". The completed scheme will restore navigation on the Forth and Clyde canal between Bowling on the River Clyde and Grangemouth on the River Forth. On the Union canal, navigation will be restored between Lochrin Basin in Edinburgh and Falkirk. A new link between the canals will be created at Falkirk, thus allowing navigation from central Glasgow to central Edinburgh and from the Forth to the Clyde. This will provide a new tourism, recreation and heritage resource, stretching the full width of the Central Belt of Scotland, which will create benefits for users and for the wider economy. The Millennium Link will create the opportunity for a series of major developments along its length - these will arise in part from the operational requirements of sailing on the canal and in part from the effect of the canal in opening up sites with latent potential for tourism and business space development.

AREA CHARACTERISTICS

1.2 Many of the economic and environmental problems of Central Scotland are the results of the "negative heritage" of past economic activity - degraded landscapes and structural economic weaknesses. The Millennium Link project will transform part of that "negative heritage", a largely derelict and severed canal, into an asset and will lead to the improvement of the environment along the 110 km length of the canals and at various sites along the canal.

1.3 Economic weakness and environmental disamenity have particularly adverse effects on the lives of people in disadvantaged areas. The Link project passes through or near many such areas and provides the opportunity to create employment and training initiatives targeted at these areas.

The Millennium Link. Thomas Telford, London, 2000, 55–71

1.4 The canals run through 60 local authority wards in nine local authority districts[1]. The socio-economic profiles of these wards are not homogenous. The wards comprise both affluent and deprived areas, as well as rural and urban areas. Substantial differences exist even between wards within a single district - for example, the Edinburgh wards include Wester Hailes which is a deprived peripheral estate and currently designated one of the 15 Scottish Office partnership areas, and Merchiston, a fairly prosperous conservation area in the centre of town. The canal wards cannot, therefore, be considered as a single group nor as nine sets of wards.

1.5 Unemployment along the canal is, unsurprisingly, concentrated in the most populated areas: Glasgow, Falkirk and Edinburgh. There is, however, considerable variation within each ward grouping. In Glasgow, Maryhill, Keppochhill and Ruchhill have the highest unemployment rates whereas wards such as Kelvindale have very low unemployment rates. In Edinburgh, Wester Hailes and Haymarket/Tollcross have the highest unemployment while Colinton and Merchiston are well below the Scottish average. Many of the affected wards are areas which are eligible for URBAN funding.

[1] The 60 wards are based on the pre-reorganisation boundaries. They are used in this section as most of the available data are produced to these geographies. It also relates to the information produced for the baseline study, produced in 1995, to inform the project bid.
It should also be noted that the number of canal wards in each district varies considerably: Dumbarton and Bearsden and Milngavie each contain just one canal ward, Cumbernauld and Kilsyth contains two, West Lothian five, Clydebank and Strathkelvin six, Edinburgh ten, Glasgow 11 and Falkirk 18.

2. PROJECT IMPACTS

2.1 The economic impact of the Millennium Link will involve:

- the creation of employment and training through the construction and operation of the Link;
- the creation of tourism, leisure and recreation opportunities;
- a stimulus to economic regeneration and development in sites and communities in a corridor stretching across Central Scotland;
- a contribution to the wider restructuring of the Scottish economy in respect of, *inter alia*, business development, new starts, inward investment sites, place competitiveness, access to opportunities and skills development.

2.2 It is important to distinguish between two fundamental categories of benefit. What have been termed the **direct** benefits arise strictly from the reconstruction and operation of the canal and do not depend upon any investment which is not specifically required for the restoration and operation of the canal. It therefore includes that part of tourism relating to informal recreation and boating activity on the canal. The term **indirect** has been used to describe those benefits which will arise from the exploitation (through further investment) of the range of development opportunities which will be generated by the canal project. As will be seen, the greater part of the employment benefits of the project will arise through the indirect route and will take several years to emerge. The project is not by any means exceptional in this respect. Many infrastructure projects which are supported by structural funds – such as site development – will not produce most, or even any, of their benefits until some time, possibly years, after the project has been completed. There is, however, already emerging evidence of partnership and developer interest, dependent on provision of the canal infrastructure.

2.3 Because some of the benefits will not emerge until some time in the future, it is necessary to give particular attention to the risks and uncertainties surrounding the project.

A: DIRECT BENEFITS

2.4 The direct benefits of the project arise in the **construction** and **operation** of the restored canal.

2.5 The works associated with the canal itself involves expenditure of £78.3 million. The Scottish Enterprise RAPID appraisal guidelines indicate that every £270,000 of such expenditure will create the equivalent of 1 full time job. On that basis, the work would create 289 FTE jobs. An alternative approach would base the calculation on the average output per worker in the construction sector. This latter calculation indicates that the project would create 1000 jobs for one year - i.e.100 FTE. The RAPID figures are used here for the "base case". Some of this employment will be generated outside Scotland since it will relate to the manufacture of equipment but the greater part of the employment will be created in the local areas. Studies of major construction projects have shown that jobs are filled mainly by people from within the local Travel to Work Area. We would expect 70 % of these jobs to be filled by people living within 30 miles of the sites.

2.6 The canal is currently classed as 'remainder waterway' and BW is bound to maintain it in the most economic way possible. After restoration, BWB will apply to have it reclassified in terms of current legislation as a 'cruising' waterway. They would therefore bear the associated higher maintenance costs. At present, BWB is able to expend additional monies in the expectation of the Link proceeding. Had the project not proceeded, this would have resulted in the canal being left as it is, with a downgrading and downsizing of British Waterways' current operation at the site and the potential loss of 40 FTE operational and maintenance jobs.

2.7 There are various initiatives which can be taken to enable opportunities to be targeted on priority groups. The government's "New Deal" for the unemployed is of particular relevance. Not all of the work to be undertaken on the canal – e.g. specialised engineering work – could lend itself to such initiatives but the main identified opportunities for initiatives to provide training and employment opportunities are as follows:

- environmental works related to the canal could provide opportunities for participation by the Environmental Task force;
- projects to enhance the environment of the improved canal may be undertaken by voluntary bodies – these could also provide placements under Welfare to Work;
- Partners are committed to develop Intermediate Labour Market projects involving improvements to the amenity of the canal (though not time critical engineering and construction work).

2.8 It is difficult to estimate the scale of the potential benefit but some guidance may be drawn from other BW canal projects. Training and employment opportunities have been provided at other BW canal projects with the support of ESF funds. Extrapolation from the experience of those other BW projects suggests that a project of the scale of the Millennium Link should provide scope for around 500 training places.

2.9 Once the canal is operational, further impacts will arise. These will involve informal and formal use of the canal. The use of the canal area for informal recreation - walks, casual sightseeing, fishing and cycling is monitored by BW using various methods including electronic pedestrian counters and survey work. From this information fairly robust estimates of recreational and informal use may be established. Using these methods, BW calculates that the central Scotland canals are used each year for over 7 million informal recreational visits.

2.10 British Waterways Board has made calculations of the level of recreational use of the canals in 1989 and 1995. Comparison of the two surveys indicates considerable growth in the number of users in the canal. This pattern of growth is reflected in other studies of the growth of recreational activity in the UK. While some growth of recreational activity is expected even without improvements to the canals being undertaken, there is evidence from other parts of the BW system that improvements to canals lead to increased levels of recreational activity. Thus improvements to the canal side at Kidderminster were followed by growth of 10% per annum in the number of visitors[2]. Similarly the improvement of the towpath of the Staffs and Worcester canal led to an increase in the number of users of 38%[3]

2.11 BW estimates that the secular growth in recreational use of canals is around 4-5% pa. This is the growth which might be expected without improvements being done. Following the examples given above, the growth in use "post improvement" could be expected to be of the order of 50% over 5 years as compared with only 25% given no improvements. The 50% growth would give an extra 3.5 million visitors. If we take the view that half of these extra visits would have arisen in any case, the "additional" visitors due to canal improvements amount to 1.75 million pa after five years.

2.12 BW survey work produces estimates of spend associated with informal recreational activities. Using the BW estimates of spend per visitor, it is calculated that the 1.75 million additional informal visitors will spend £6.0 million pa. Thus although the number of informal visitors is large, their spending, as one might expect, is low. This **additional** spend would, on the basis of research on Tourism Multipliers[4], produce 238 FTE jobs, divided as follows:

[2] British Waterways Towpath Improvement Kidderminster - Impact on Visitor Satisfaction and Numbers. 1995
[3] British Waterways Towpath Improvements Bratch Locks - Impact on Visitor Satisfaction and Numbers. 1995

West of Scotland	135
Eastern Scotland	74
Edinburgh	29

2.13 This recreational activity is all additional but it will involve some degree of displacement. That is, if the additional trips had not been made some of the spending involved would have otherwise been spent on other goods or services in the local or national economy. The recreational activities will be substitutes for other, similar, activities which would otherwise have been undertaken locally or elsewhere. Since this is mainly activity undertaken on short trips from home, displacement at the national level must he high – of the order of 90%. This has the consequence that Scottish level job creation would be about 24.

2.14 The restoration of the canal system will lead to entirely new tourism activity involving cruising on the canal. Canal-based tourism activity will consist of:

- "Transit" craft moving from one coast to another;
- Boats kept on the canal system by private owners;
- Craft hired by holidaymakers;
- Boats offering passenger trips, including restaurant boats.

2.15 The forecasts of water borne demand are drawn from the document **The Millennium Link: A Demand Analysis for Leisure and Tourism Activities on the Re-Opening of the Forth and Clyde and Union Canals** (British Waterways, July 1996.) This report covers both towpath users and boating traffic. The estimates of boating traffic in the document are drawn directly from an earlier (1995) report prepared by ASH and Coopers and Lybrand entitled **Millennium Link: Tourism Study**.

2.16 The following assessments have been made about the level of **additional** boating use which is expected to be achieved within the first five years of re-opening.:

• yachts and other craft in transit between east and west coasts:	500 trips p.a.
• moored private boats:	600
• moored hire boats	40-60
• trip boats	6
• restaurant boats	2
• unpowered boats	14,000

[4] Surrey Research Group **Scottish Tourism Multiplier** Study 1992

2.17 The levels of spending associated with these additional activities are estimated to be as follows.

Source	**Spend pa (1996 prices)**
Transit Boats	£60,000
Moored Private Boots	£1,585,000
Moored Hire Boats	£1,000,000
Trip Boats	£155,000
Restaurant Boats	£300,000
Unpowered Boats	£40,000
Total	**£3,140,000**

2.18 Using the same figures for tourism spend per job as above, it is estimated that this total spend would generate 120 jobs. The expected distribution of these jobs is as follows:

West of Scotland	67
Eastern Scotland	38
Edinburgh	15

2.19 The development of this new tourism market may give rise to some displacement. However, the opportunity to transit Central Scotland by canal will be entirely new and is unlikely to displace other sailing activity in Scotland. Anyone who wants to undertake a canal narrow boat cruising holiday at present would have to go to England or Wales - the Caledonian canal is quite different in character from the Millennium Link. Any displacement would, therefore, be from holiday activities which did not involve sailing or boating. Given this, displacement is likely to be very low - indeed, it is arguable that displacement would be close to zero given the absence of alternative similar facilities in Scotland and the low level of such activity at present (sailing/boating does not feature in the STB's list of activities undertaken by visitors to Scotland). However, as a conservative estimate, displacement of 20% is adopted.

2.20 The operation and maintenance of the canal will also safeguard approximately 40 British Waterways jobs across the canal area

2.21 Table 1 sets out a summary of the costs and benefits directly generated by the proposed project. The **temporary** jobs and the **training places** will be achieved within the canal construction period - i.e. by 2001. The **User Benefits**, and permanent jobs will be achieved within the first five years of operation - i.e. by 2006. The issue of timing of benefits is reconsidered below. This assessment of direct benefits takes no account of the very substantial indirect or catalytic effects discussed below.

Table 1

Millennium Link – Net Direct Costs and Benefits

Costs	**Benefits**				
FINANCIAL: Construction - £78.3 million Net Additional Running Costs - Nil	**Benefits**	**Scotland**	**West of Scotland**	**Eastern Scotland**	**Edinburgh**
	Jobs FTE **Informal Recreation**	24 (90% displacement assumed)	66 (50% displacement assumed)	36 (50% displacement assumed)	17 (50% displacement assumed)
	Jobs FTE **User Spend**	96	54	30	12
	Jobs FTE **Maintenance**	40	22	12	6
	Const. Jobs (FTE)	289	81	150	58
	Training Places	250	70	130	50

Additional Benefits - These are qualitative and would include:

- Improved quality of life
- Improved social & economic inclusion
- Sustainability
- Contributing to restructuring of the Scottish economy through business development, new starts, providing inward investment sites, place competitiveness, access to opportunities and skills development

B: LONGER TERM BENEFITS

2.22 The indirect, longer term benefits will arise from the development of a series of sites across Central Scotland. These benefits arise from two distinct areas of economic activity but will often be found together on the same site.

- The first element of benefit relates to tourism. The Link will create a corridor of tourism opportunities as certain key nodes on the canal route afford the opportunity to develop significant visitor attractions based on the drawing power of a working canal and the canal's diverse heritage.

- The provision of property to meet the needs of modern, 'footloose' business. These benefits will be felt in terms of employment creation and the expansion of small and medium sized enterprises which will form part of the market for the new sites.

Canal-Related Tourism Development

2.23 In conjunction with the project partners, the DTZ Pieda Consulting study team which worked on the Millennium Link development study identified a number of important sites capable of canal- related development within a five year time horizon Following analysis of the potential, it is considered that 4-5 could be developed as significant visitor attractions with the capacity to attract visitors from local, national and external markets. The other sites would be secondary attractions. Possible commercial and tourism related activities have been identified for those sites. The gross impact of this activity has been analysed in two ways - by assessing the level of employment likely to be associated with the envisaged scale of development and by assessing the potential for tourist spend.

2.24 The sites have been assessed primarily as "non paying" attractions. That is, while there would be on-site activities which would earn revenue, there would be no charge as such for access to the sites. Scottish Tourist Board data indicate that there are around 40,000,000[5] visits made each year to tourist attractions in Scotland. The number of attractions drawing over 200,000 visitors pa is around 30. Second rank attractions such as smaller Country Park visitor centres attract 50,000 - 100,000 visitors pa. The study team consider that 5 of the sites could be developed to attract up to 200,000 visitors pa while average visits to the other 15 sites would be able to attract 50,000 - 70,000 each. This gives an approximate total of 2,000,000 visits. This represents a growth in the market of 5% over five years - given that, according to STB data[6], visits to attractions grew by 8% between 1995 and 1996, the target growth does not appear excessive.

---

[5] Scottish Tourist Board Tourism in Scotland 1996

[6] op cit

2.25 Scottish Tourist Board data indicate that the mean spend per day by tourists is around £36, of which about half would be on accommodation. Home based day visitors spend much less - about £8 per trip. Given the likely mix of visitors (30% non Scottish), a reasonable average spend associated with each visit would be £12. On that basis, 2,000,000 visits would produce spend of £24,000,000 - sufficient to support 930 jobs. These are conservative estimates - spending levels in particular may be higher given that the attractions will be well supplied with facilities and in urban areas. Moreover, Tourist Board data[7] indicates that 20% of visitors to attractions in Scotland are from overseas, with a further proportion being non Scottish (there are twice as many English tourists as foreign tourists in Scotland). Foreign tourists typically have higher spending per day than domestic visitors. The analysis of the site development profile and the pattern of likely development (shown in Appendix 3) suggests that the sites could accommodate around 1,100 jobs in tourism activities.

2.26 The extent to which this activity would be displaced from other tourism in Scotland must be considered. As has been noted, the proposed developments would essentially accommodate growth projected on the basis of recent trends and may thus prevent the emergence of frustrated demand. A large proportion of visitors, as much as 40% could be non Scottish. It is known that non Scottish tourists concentrate on the larger attractions such as historic properties so we may assume a lower non Scottish share at 30%. If non Scottish visits are largely non displaced while the reverse is true of Scottish visits, we may estimate displacement at the Scottish level to be 50% while at the individual partnership area level it would be lower - since domestic visitors would come in part from outside the area. Displacement of 30% at the local level has been assumed.

Tourism Related Training Opportunities

2.27 At this stage it is considered that 500 one-year training places in total could be provided on construction work related to new tourist development (i.e. over and above the work on canal restoration). The scale of this potential is based on the scale of the spending and comparison with the canal restoration work described above. A further consideration is that this work is some way off - most will not begin until after the turn of the century so that the opportunity to plan ahead and develop initiatives using the Welfare to Work mechanism and ESF resources is considerable.

---

[7] BTA Research Services Sightseeing in the UK 1994

Community Business

2.28 Similarly, the Community Business sector could become involved in meeting the market of 2,000,000 visitors predicted. This opportunity could be exploited if a positive strategy was implemented in the priority areas to identify business possibilities and to promote community business. Estimates of impact at this early stage must be speculative but the development of 20 community businesses across Central Scotland, employing say 100 people, would be a realistic target.

SMEs

2.29 It is in the nature of tourism and leisure than may enterprises involved in the sector are small. The consultants have reviewed the potential development at each of the sites and consider that the various tourist businesses could credibly involve 50 SMEs spread along the corridor. These would range from shops to chandlers serving the boating market.

2.30 Additionally, the commercial site development will certainly accommodate SMEs. The study team would expect, on the basis of the expected scale of development and the typical profile of business site occupancy in Central Scotland that 60 SMEs would be accommodated on the business sites.

High Amenity Business Sites/Industrial Development

2.31 An opportunity afforded by the canal to develop high amenity business sites. The importance of providing such sites and the lack of such sites in certain areas, notably Glasgow, are widely recognised. Although the property needs of economic activity are diverse, the major growth area for demand in recent years has been in relation to business park space and office space - reflecting the continued restructuring of the economy and the changing nature of production processes. It is also the case that the proportion of businesses seeking a prestigious or distinctive location is growing.

2.32 Scotland has developed a number of successful business parks in recent years but the canal affords the opportunity to add to this development. This is not a case of the canal creating the demand but rather releasing the supply potential. The study team consider that there are at least four major development sites along the canal corridor which would be released or enhanced by the canal project. That is, these are sites which have potential and which are located in areas of demand potential but where quality development, or even any development, may be difficult to secure without the impetus of the canal project.

2.33 The canal project will focus demand on certain sites, enabling investment to be supported while the enhanced environment of the restored canal will stimulate interest in the sites concerned. The precise location of the development sites is open to discussion but the general market conditions are such that the potential for development will be greatest in Glasgow, in the Kirkintilloch area, in Falkirk/Grangemouth and in West Lothian and Edinburgh.

2.34 The employment associated with all industrial and business development of the sites has been calculated at just over 5,000. This figure has been adjusted for expected levels of displacement (50%). A multiplier of 1.2 has been applied to the net additional jobs and the results are shown below:

Industrial and Commercial Impacts		
Area	**Net Jobs**	**Incomes**
West of Scotland	1800	£36.0 million
Eastern Scotland	792	£16.0 million
Edinburgh	200	£4.0 million
Total	**2792**	**£56 million**

Construction Employment

2.35 The site development works would also create a benefit in terms of employment. The impact of the capital spending associated with the above developments, including site remediation work, is expected to provide, using the RAPID guidelines mentioned earlier, 444 FTE jobs.

C: Summary of Total Project Impacts

2.36 The overall economic impacts of the project are shown in Table 3 along with the related activities which will generate the impact.

2.37 The timing of the benefits is straightforward. The direct benefits will be procured over the period to 2001 with the construction jobs benefits accruing as soon as work begins. The indirect benefits shown should be obtained within 5 years after opening. There may be further benefits in the long term.

Table 3
Millennium Link - Summary of Benefits to Scotland

Action	Cost	Intermediate Output	FTE Jobs	Const. Jobs FTE	New SMEs	Other Benefits
Direct Benefits						
Canal Restoration	£78.3 million	Restoration to Navigability of Forth and Clyde/Union Canals. Creation of construction employment		289		250 Training Places.
		Increase in Informal Recreational visits to Canal by 1.75 million pa	24			
Canal Operations	Net Cost - nil	Use of canal by various cruising and transit craft. Increased tourist/user spending of £3.1 million pa. Protection of operational employment	136			
Indirect Benefits						
Creation of new canal-related tourist attractions at various sites along the canal corridor.	Site preparation costs of at least £20 million. Private sector investment of around £100 million	Creation of 4/5 major and around 20 secondary sites attracting annual total visits of about 2 million and tourist/visitor spend of £32 million	460	444	50	250 Training Places. 20 Community Businesses employing 100 people
Creation of new business parks at sites "opened up" by canal project		Development of 4 major business sites and 4 other sites containing business premises	2800		60	
TOTAL			**3420**	**733**	**110**	

Table 4
Sensitivity Analysis

Benefit	Key Variable	High Variant	Low Variant
Construction Employment – 544	£ per FTE job	None	£800,000 per FTE - 184
Employment Created by Canal Users – 136	Number of Users	Numbers 50% Higher - 204	Numbers 50% Lower - 68
Indirect Tourism Jobs – 465	Number of Visitors	2,500,000 visitors - 580	1,000,000 visitors - 232
	Displacement Level	None	70% Displacement - 280
Business Jobs	Number of Developments	None	Two Main Sites only - 1400 jobs
	Displacement	None	70% Displacement - 1675

3. SENSITIVITY ANALYSIS

3.1 Results of the analysis are sensitive to a range of parameter values. Some variations have, however, relatively little impact on the outcomes - i.e. the employment impacts are not sensitive to variations in the level of informal recreation. It is possible to set up various permutations of variables to produce different mixes of results. The approach taken here is to consider each key output in turn and consider which variables are most likely to impact on that output.

3.2 The results of this process are shown in Table 4 overleaf which, in relation to each output, identifies the variables to which the results are most sensitive and indicates the impact of alternative values for these variables on the element of output.

4. CONCLUSION

4.1 This report has presented the socio-economic impact analysis for the Millennium Link. The project represents a major partnership development with long term benefits to Scotland that will assist in the achievement of partners' shared strategic and operational priorities. The project will transform a large area of central Scotland and generate a wide range of positive and substantial benefits covering regeneration, tourism, social inclusion and wider market and community effects.

4.2 The project produces a mixture of benefits. A summary of the project benefits is provided in Table 5 below

Table 5: MILLENNIUM LINK - SUMMARY OF ECONOMIC BENEFITS	
DIRECT – CONSTRUCTION AND OPERATION OF THE RESTORED CANAL	
REGENERATION	**£78.381 million investment**
	• 120 ha of canal/adjacent land benefiting from environmental improvements • 290 FTE construction jobs • 70% plus of employment within a 30 mile radius of the site • 500 construction training places • 40 FTE operation and maintenance jobs safeguarded
TOURISM	**1.75m additional visitors pa due to canal improvements**
	• £6m pa of spend associated with informal visitors • 25 FTEs at the Scottish level • £3.1m pa of spend associated with additional boating use • 100 FTEs at the Scottish level
SOCIAL INCLUSION	**Opportunity to create employment and training initiatives targeted at disadvantaged areas experiencing economic exclusion and environmental disamenity.**
	• Environmental Task Force • Voluntary Sector • Intermediate Labour Market • Community Economic Development Linkages • Sustainable Development
INDIRECT – DEVELOPMENT OF SITES ACROSS CENTRAL SCOTLAND	
TOURISM	**The Link will create a corridor of tourism opportunities at key nodes along the canal.**
	• 5 significant visitor attractions (200,000 visitors pa) • 15 smaller attractions (50,000 – 70,000 visitors pa) • 2 m visits pa producing spend of £24m pa • 465 FTEs at the Scottish level • 50 tourism SMEs along the corridor • 20 community businesses employing 100 FTEs
BUSINESS SITES	**Major development sites along the canal corridor released or enhanced by the project.**
	• 4 major development sites identified, with potential, in areas of market demand • Enhanced environment of the restored canal stimulating developer interest within 5 years • Areas of potential include Glasgow; Kirkintilloch; Falkirk/Grangemouth & West Lothian and Edinburgh • Potential to accommodate 2800 net FTEs and generate income of £56m • Interest from housing developers based on the unique opportunity afforded by the canal

The Millennium Link: Falkirk case study

STEPHEN DUNLOP
Director of Community and Environmental Services, Falkirk Council, Falkirk, Scotland.

INTRODUCTION

The Falkirk Council area is at the heart of the Millennium Link, taking in substantial stretches of both the Forth & Clyde Canal and the Union Canal, including the Wheel interchange and the connection with the tidal waters of the River Forth.

This Falkirk Case Study examines the Millennium Link from the perspective of a local authority. It describes the Council's aspirations for the project in its area, the processes it has engaged in and some of the issues it faces in realising the vision.

THE FALKIRK COUNCIL AREA

The Falkirk Council area covers a compact area of some 300 sq. kms. on the south side of the River Forth in central Scotland, equidistant from Edinburgh and Glasgow, with a population of 143,000.

Map 1 : Forth & Clyde Canal & Union Canal

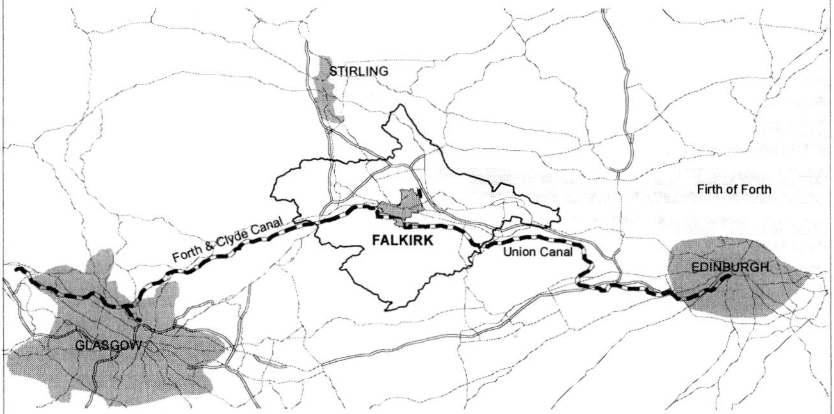

It can boast that its position has been of strategic importance for defence, communications and trade for at least two millennia. In the construction of the Forth & Clyde Canal across the neck of land between the estuaries of the two rivers, the 18th Century canal builders followed the same route as the Romans some 1600 years earlier who built the Antonine Wall as a fortification for occupied Britain to the south. Central Scotland's topography and its centres of population also dictated that, when railways and motorways came to be built in the 19th and 20th Centuries, they would follow the same east-west axis which, in its eastern section, traverses what is now the Falkirk Council area.

The Millennium Link. Thomas Telford, London, 2000, 72–78

The character of Falkirk's development has depended not on its location alone but on its intrinsic industries. As a leader and major centre of ironfounding and coal mining industries from the 18th Century onwards, with the development of engineering industries and with a modern petrochemical and port complex at Grangemouth today, much, though by no means all, of the area is or has been industrialised and, in the early period, the canals enabled and accelerated that process of industrialisation There still remain significant rural areas and both the Union Canal on its route east towards Linlithgow and the Forth & Clyde Canal west of the Wheel site pass through areas of attractive countryside.

There is a price to pay for the benefits of industrialisation when that industry is no longer needed and the concentration of industry in Falkirk in the past has meant that the area has inevitably suffered severely from the decline in traditional industries over the last 30 years. The environmental, economic and social effects are clearly seen in factors such as derelict land, redundant buildings, unemployment and social exclusion. These are factors which Falkirk Council, since its formation in 1996, and its predecessors have been grappling with for some time and not without significant success. A regenerated town centre and an unemployment rate falling in line with the Scottish average indicate that restructuring has been and is taking place in the local economy. However, there is still a desire, and a need, to move that process along faster.

THE MILLENNIUM LINK AND FALKIRK COUNCIL'S ASPIRATIONS

It was against this background that Falkirk Council began to form its views on the Millennium Link proposals. At one level, they represented a large 'windfall' investment in the area and as such had to be welcome. It was estimated that approximately half the total project funding of £78m would be spent on infrastructure in the Council's area. In an era of seemingly ever more control on local authority capital spending, the capacity of the Council alone to invest in order to move the economy forward is quite limited.

Map 2 : Context

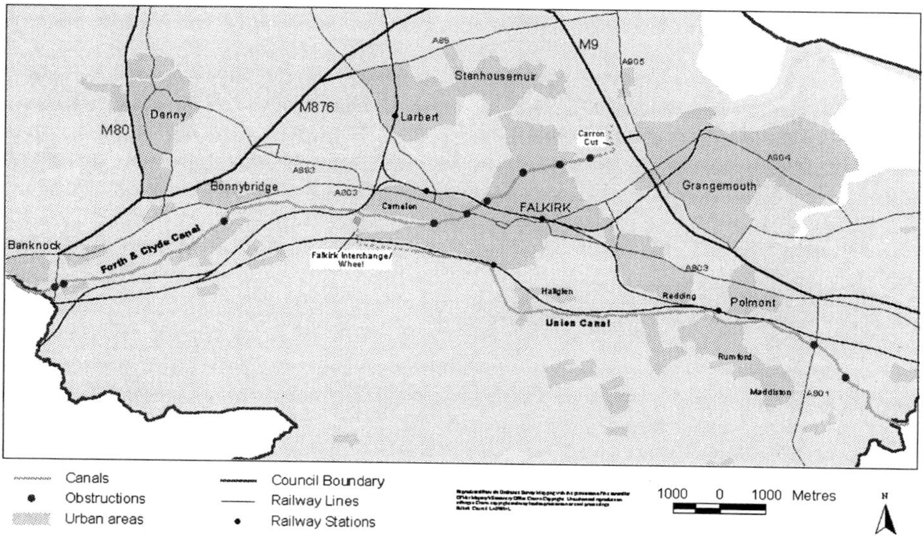

At the same time as welcoming the investment, some people may, at least inwardly, have questioned the wisdom of rebuilding an 18th Century mode of transport, maximum speed 4 mph, in the 21st Century. Was this really where Falkirk's future lay?

Well, after some further thought, the answer seems to be yes, at least part of Falkirk's future lies in the project. Modern industry and the information technology revolution must of course play a large part in generating the jobs and income of the future but the restoration of the canals is clearly seen as offering new potential for the tourism, leisure and heritage industries and these represent real economic and environmental opportunities. Furthermore, by cultivating diversity in the local economy it may gain a strength and resilience which it has sometimes lacked in the past.

Investment in leisure and heritage is of course nothing new for local authorities and Falkirk Council has been particularly active in recent years in projects such as the restoration of Callendar House, Grangemouth Sports Stadium and Bo'ness and Kinneil Railway. The area is now regarded as an emerging player in the leisure tourism industry and has also seen much investment in its tourist accommodation stock, primarily for the business market.

Part of the attractiveness to the Council of the Millennium Link is that it is consistent with the existing strategy of building up the tourism market. Developing new canal-related facilities will generate additional visitors to the area and the spin-off should boost return on other investments already made or committed. Another welcome feature is that as the canals pass through many different areas and communities of the Council, the benefits need not be confined to one or only a few locations. However, it is also recognised that the Millennium Link has the potential to transform the scale of the tourism industry and its economic impact in the area. It is as a fast forward route to developing leisure, heritage and tourism facilities, including the creation of a dramatic centrepiece in the Wheel, that the Millennium Link really begins to excite Falkirk Council. The high profile of the project gives the Council an opportunity to enhance the image of its area, turning what is arguably its negative, industry-related image in Scotland, and perhaps its non-image further afield, into a much more positive and widespread understanding of Falkirk as a location. Perhaps the Wheel can be the 'icon' which rebrands Falkirk.

Along with the excitement generated, it has been recognised that much of the economic potential of the Millennium link will be realised only if the Council, other agencies and development partners engage in a process of planning, co-operation and promotion in order to facilitate development and that this process will require significant additional investment. Furthermore, although the project celebrates the beginning of a new Millennium, whether that is regarded as being in January 2000 or 2001 it is really only the start date for the economic development phase of the project, which can be expected to last over perhaps a 10 or even 20 year period. In embarking on this project, the Council has come to realise that it is likely to be in it for the long-term and this will have to be taken into account in determining future priorities and the allocation of resources.

THE PROCESSES AND THE ISSUES TO DATE

As a local authority, Falkirk Council has statutory powers, such as planning, and discretionary powers such as economic development. At the beginning of the Millennium Link project, the Council already had a range of policies relevant to the development of the canal corridor. These included the statutory Structure Plan and Local Plans (including the Forth & Clyde Local Plan) and the non-statutory Economic Development Strategy, the Environmental Strategy, the Culture and Leisure Strategy and the Community Urban Regeneration Strategy.

The Millennium Link can be regarded as having two main aspects. One is the engineering works to re-open the canal, which involve the Council in a number of statutory responsibilities and as a funder. The other aspect is the economic development opportunities in the canal corridor, in which the Council also has a major interest and role, mostly through the exercising its discretionary powers.

Engineering

To facilitate British Waterways' construction programme, the Council has had to negotiate and handle land transfers and to deal with many planning applications and road traffic issues. A Millennium Link Corporate Working Group was set up within the Council to bring personnel together from different Council services in order to expedite British Waterways' business, as far as possible to meet an often tight construction timetable.

However strong the Council's commitment to the project, it still has to act in a way which protects its own interests and discharges its responsibilities in accordance with statutory procedures. Inevitably, there have been occasions when time pressure and tension have built up. There was also disappointment in the Council at some of the engineering decisions, in particular the decision to route the canal into the River Carron, rather than to the original terminus at Grangemouth Docks. It was hoped that the latter option, as well as re-integrating the town of Grangemouth with what was regarded as 'its' canal, would have acted as a catalyst to promote the much wider regeneration of disused dock areas. While the Grangemouth decision was a disappointment, other decisions, such as the re-design of the Wheel have been well received.

In supporting the engineering works, it has surely been beneficial that one organisation, the Council, has had the powers to deal with a wide range of different issues and, because the Corporate Working Group effectively gave British Waterways a 'one stop shop' for Council services, Falkirk Council considers that it has been able to provide effective support for this phase of the project. The Council's contribution to the engineering works has also been boosted, sometimes fortuitously, by its own construction programmes, such as the North Orbital Road which routes industrial and through traffic away from the canal, allowing a bridge obstruction to be removed.

Economic Development

At an early date, there was some appreciation of the general types of opportunities presented by the Millennium Link but the actual sites and conditions likely to influence development were less well understood. In 1997 the Council set up a group, including representation from Forth Valley Enterprise and British Waterways, to undertake a Canal Corridor Development Study. The outcome has been the production of the Canal Corridor Development Framework, including an opportunities appraisal of 10 key sites in the Falkirk Council area. The study was carried out in parallel with a DTZ Pieda study but looked in greater detail at sites within the Falkirk area. The Framework document incorporates the Council's vision with regard to issues of sustainability and environmental protection as well as development and 'spin-offs' in the wider economy.

MAP 3 : Opportunities

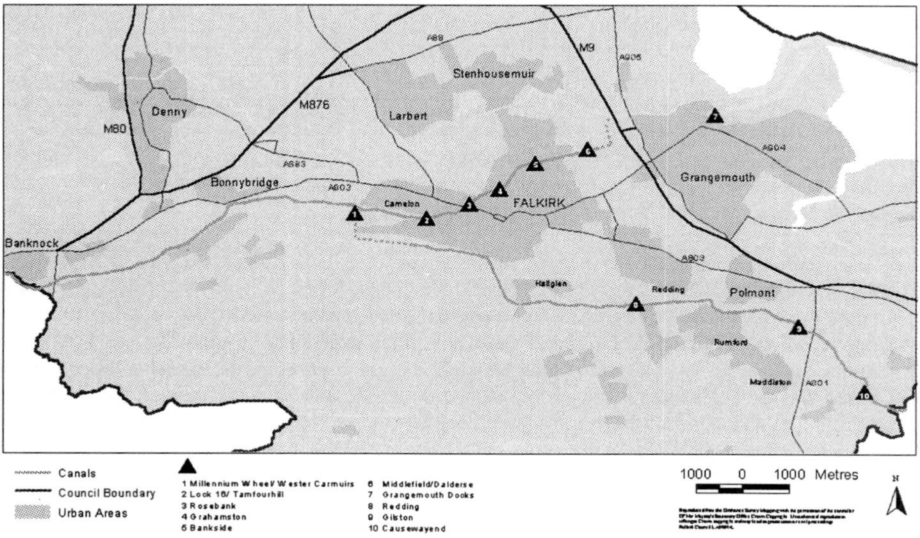

As well as identifying opportunities, the Development Framework identifies constraints. Some constraints, such as land contamination on strategic sites and difficult access issues, may only be overcome by high development costs and in such cases a substantial scale of development may be required to generate the necessary return. However, other constraints limit the scale, and perhaps the location, of development, such as the requirements of the Health & Safety Executive at sites in Grangemouth and protection of the archaeological and natural heritage. These are matters for developers and the Council to address case by case.

The Framework is not a blueprint for development nor a piece of marketing literature but is intended to give guidance to developers and the Council alike, set an agenda for the Council's own proactive efforts to regenerate the canal corridor and assist the Council in responding to canal-related proposals from the private sector. There is already evidence of development pressure at places within the corridor, though not all proposals have been on suitable sites or have been considered appropriate. As well as encouraging development, the Council must be in a position to protect sites for other purposes and longer term economic development.

The study group, with additional representation from Scottish Enterprise, has evolved as the Millennium Link Development Group, providing a vehicle for promoting development opportunities and co-ordinating work carried out by the various partners.

The Falkirk Wheel

Clearly, the Falkirk Wheel will be the most significant feature on the canal, not only in terms of engineering but also in terms of economic development opportunities. The Wheel will be a unique and high profile visitor attraction in its own right and its ability to draw additional visitors to the area will create a market for possible additional facilities such as hotels, restaurants, boat and cycle hire, speciality shopping, interpretation and entertainment facilities and perhaps a holiday park, sports facilities and conference centre. As the Wheel is being constructed on an open site, there is the prospect of building up the range of such facilities on one large site in order to maximise the appeal of the location and its economic impact.

However, the Wheel site and surrounding area not only have great potential but also have a range of major problems affecting future development. These include contaminated and difficult ground conditions, land assembly issues, the need for a new road access and the presence of archaeological remains. While most constraints can, technically, be overcome, any development would have to cover the associated costs.

The results of a marketing exercise to potential developers and a consultancy study commissioned by Falkirk Council and partners appear to indicate that there is still a degree of uncertainty in the market as to the viability of commercial ventures at the Wheel site. One major constraint is the cost of an alternative access road which would be required for an expanded range of facilities. Such a road would have the benefit of approaching the site from the north, a much more convenient and direct route for visitors, but at an estimated cost of £3m.

The Council is determined that the Millennium Link and the Wheel Site in particular should deliver appropriate economic benefits. Although certain types of major development, such as housing or retailing, could fund the required infrastructure costs, they would also displace the potential development of facilities which bring visitors and therefore would limit future economic benefits, as well as being contrary to planning policy. Getting the right sort of development in the longer term is much more important than getting any sort of development now. At present the Council is planning, through the Transport Challenge Fund, to develop a shoppers' Park and Ride facility to the north of the Wheel site and this would involve the construction of at least the first section of the proposed new access road to the Wheel site, as well as providing car and bus parking facilities for Wheel visitors.

It is expected that, as the Falkirk Wheel rises from the ground over the coming months, interest from all quarters will increase sharply and potential developers will be able to envisage and assess the site's potential more accurately. It should also be recognised that the development is likely to be phased over a period of years.

CONCLUSION

Falkirk Council's experience of the Millennium Link to date has brought a lot of factors to the fore. As a project of national and international significance it has created excitement locally and has offered a new opportunity to re-profile the Falkirk area, enhance its image and promote it as a location. Although it is a British Waterways project, it also involves many partners in funding and facilitating the development and events have shown that partnership makes great demands, in terms of communication, the building of trust and recognising that partners have different and legitimate expectations from a project. Genuine partnership is difficult to achieve. As a catalyst for promoting private sector development, the project has undoubtedly stimulated interest but has also shown that there is a degree of market wariness. Further planning, pump-priming and marketing input is likely to be required from the public sector over a period of years in order to build up the right mix of development and to maximise the economic return to the area.

The Millennium Link project is a great and indeed a fitting opportunity for the Falkirk area. The canals in the 18th and 19th Centuries did much to speed up the early process of industrial development and make the area an industrial centre of international renown. At the beginning of the 21st Century, the tourism leisure and heritage industries are at a relatively early stage of development locally but the canals, once regenerated, will again have the potential to speed up the development process and, hopefully, to contribute as much to the future economic fortune of the area as they have in the past.

New Solutions

GEORGE BALLINGER
B.S.c.(Hons), C.Eng., M.Sc., M.I.Struct.E.

INTRODUCTION

The restoration of the Lowland Canals required innovative solutions to seemingly insurmountable engineering challenges. Three motorways and many major roads now criss cross the line of the original waterways. The crossings themselves have been used as conduits for public utilities to traverse the route often under cover of the Public Utility Street Works Act.

Various feasibility studies have been commissioned since a reawakened interest surfaced on the recreational and environmental significance of the canal corridors. Closure in 1963 for the Forth & Clyde and 1965 for the Edinburgh and Glasgow Union Canal was, within a mere 10 years, seen to have been short-sighted. However, that 10 year period saw the roads authorities take full use of their new found powers and old bridges were demolished almost at will to make way for wider roads with geometry that allowed for ever faster travel by motor vehicles. New trunk roads and motorways such as the M8, A80 and M9 were all constructed during this time and the asset that fed the industrial revolution tumbled into an abyss of decay and dereliction. Then, gradually, the voluntary sector began to have their voices heard and even British Waterways sat up and took notice of the built and natural heritage which had developed over two hundred years and had so nearly been destroyed in ten.

Small scale projects such as the Glasgow Canal Project, the West Lothian Canal Project and the replacement of many of the rotting timber lock gates with new steel ones began to show what could be done if limited headroom could be accepted.

THE OBSTRUCTIONS

Of course, the first thing to establish was the extent of the problem. Success required the removal of all obstructions – not 99%! Conceptual solutions were established and their feasibility tested.

Table 1 shows a list of the obstructions and the solutions are grouped under the following headings:-

1. Dropped Below Canal.
2. New Road on New Line.
3. New Road on Existing Line.
4. Canal Lowered with new Lock.
5. Opening Bridge.
6. New Canal.
7. Drop Lock.
8. New Bridge.
9. Bridge Raised.
10. New Bridge & Lock.

The Millennium Link. Thomas Telford, London, 2000, 79–85

FORTH & CLYDE CANAL
Solution

Obstruction	Dropped Below Canal	New Road		Canal Lowered With new lock	Opening Bridge	New Canal	Drop Lock	New Bridge	Bridge Raised	New Bridge & Lock
		On New Line	On Exist. Line							
Erskine Ferry Bridge					X					
Dalmuir Pipes							X			
Kilbowie Road		X						X		
Mall Pedestrian Footbridges					X					
Argyle Road			X					X		
Duntreath Avenue								X		
Clobberhill Pipes		X				X		X		
Cleveden Road			X					X		
Balmuildy Road			X					X		
Cadder									X	
Hillhead Road					X					
Auchendavie 54" Water Main	X									
Twechar					X					
Auchinstarry		X						X		
Wyndford		X						X		
A80			X					X		
Castlecary Bridge			X					X		
Bonnybridge					X					
Lock 16		X						X		
Camelon Road				X				X		
Marshalls Bridge New Locks & Bridge										X
Bainsford Road				X				X		
Lock 4		X						X		
Grangemouth Sea Link						X				

EDINBURGH & GLASGOW UNION CANAL

Obstruction	Dropped Below Canal	New Road & Bridge		Canal Lowered	Opening Bridge	New Canal	Drop Lock	New Bridge	Bridge Raised	New Bridge & Lock
		On New Line	On Exist. Line							
A801			X			X		X		
Vellore Road			X							
Greendykes Road			X							
M8						X		X		
Wester Hailes		X	X			X		X		
Kingsknowe Road			X					X		
Leamington					X					

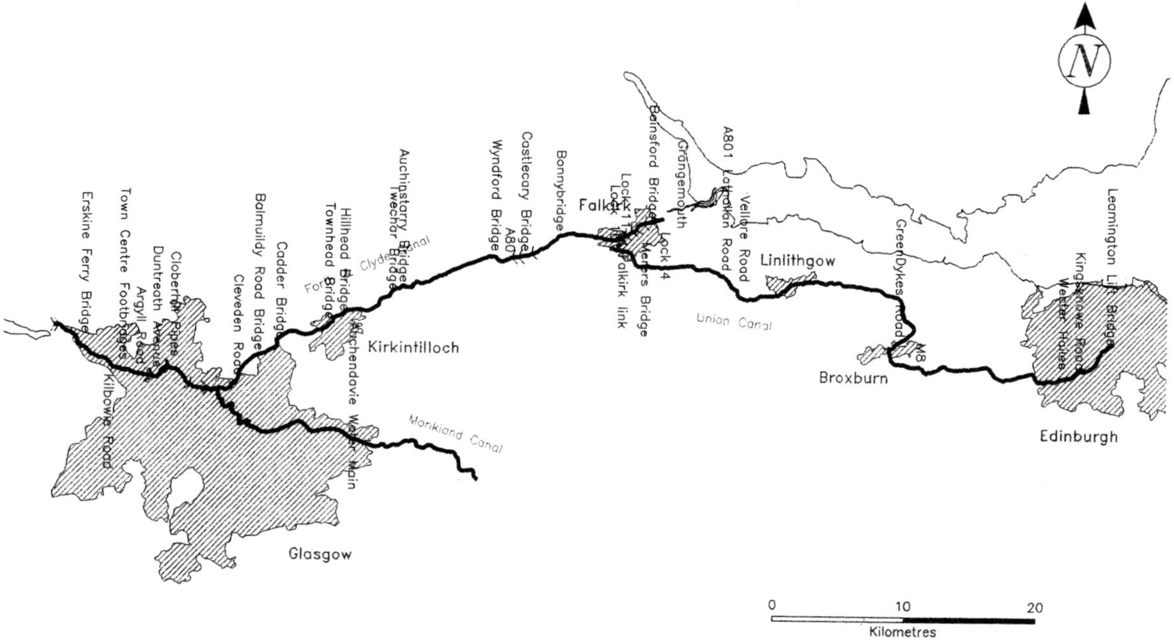

DESIGN PARAMETERS

The navigational envelope had to be established for the new canals, taking account of potential use, previous solutions and practical new solutions. What is meant by this? A few examples will show the thought process.

1. **M8 Motorway (Union Canal)**
   The original canal level was only about 1m below the road level and it was impractical to lift the road at this point because of the length that would be affected. The geometry of the road meant that headroom would be gained if the canal were diverted westwards. The limit to a westward diversion was the location of a 750mm diameter high pressure gas main which travelled under the motorway at this point. Diversion of the pipe was feasible but very expensive. Exclusion zones were established for the pipe as were deck thicknesses for the proposed bridge. This led to a 9 ft headroom from water level to bridge soffit. This fitted reasonably well with the envisaged working requirements and gave sufficient headroom at the towpath for cyclists and walkers to travel with ease under this main arterial blockage.

2. **Cleveden Road (Forth & Clyde Canal)**
   The road alignment at this crossing is substandard but any change, even an improvement, would be unacceptable unless it fully met all the modern highway standards.

   The most practical solution was to maintain the existing road geometry and dredge the canal to its original base within the pound between locks 25 and 26. Hence a distance between bridge soffit and canal base of 5.3 m was defined. Other bridges had been built to allow 3m between water level and bridge soffit and hence a water depth of 2.3m was the maximum that was practically possible, despite the original canal being 2.85m deep. This 2.3m depth was then used for all new structures eg; locks and bridges.

THE SOLUTIONS

All the solutions have had to be developed in parallel in order to ensure the 110kms of waterway truly linked together. The Grangemouth section and the Falkirk Link are dealt with in other papers and this paper will deal with two of the most novel solutions elsewhere on the project. It is a fact that there is some novelty value with all the obstructions but Dalmuir Pipes and Bonnybridge culverts are two of the finest examples.

Dalmuir Pipes

This low level road/canal intersection in a residential area with a hospital access road was originally the site of a bascule, latterly swing, bridge.

Today, however, an opening bridge is not feasible and the requirement to raise the road 3m would adversely affect adjacent housing and business premises.

A new solution was required and a "drop lock" (see sketch 1) was the most practical solution. The top sketch shows the normal arrangement with the lower sketch showing a craft navigating under the obstruction with both pairs of gates in the closed position.

There are obviously inherent dangers with such a solution. A barrier is required to prevent a boat drifting under the bridge during the filling operation and a further barrier is required to prevent impact on the gates during operation. Sudden inundated must be guarded against and the lock empties by gravity feed to the nearby Duntocher Burn. An alternative pumped system is also included. An additional safeguard is the provision of a walkway under the bridge so that a member of staff is dealing with any emergencies from a "land" base with the possibility of boats being "walked" through with no requirement for a crew or power. To ensure the head balance is maintained during operation a balancing by-pass pipe runs parallel to the lock chamber. Sketch 2 shows a longitudinal section through the drop lock with the safety features marked in red. A photograph showing the location is attached to sketch 2.

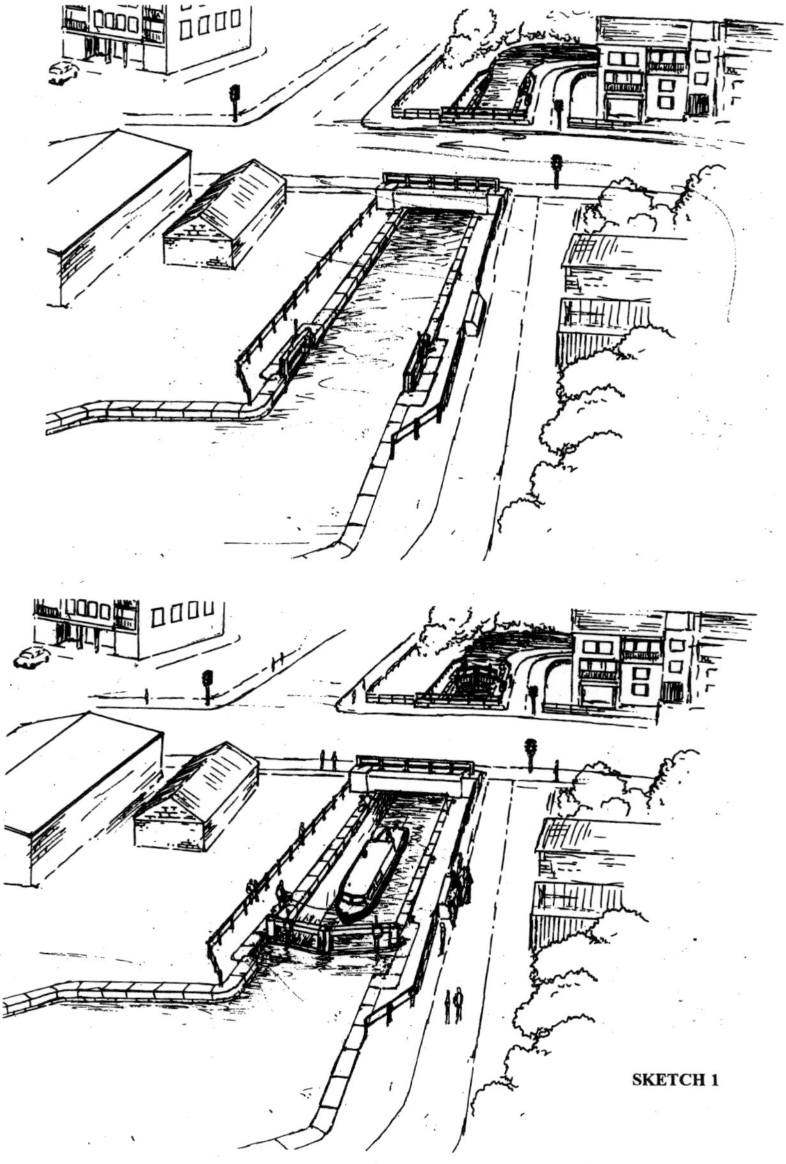

SKETCH 1

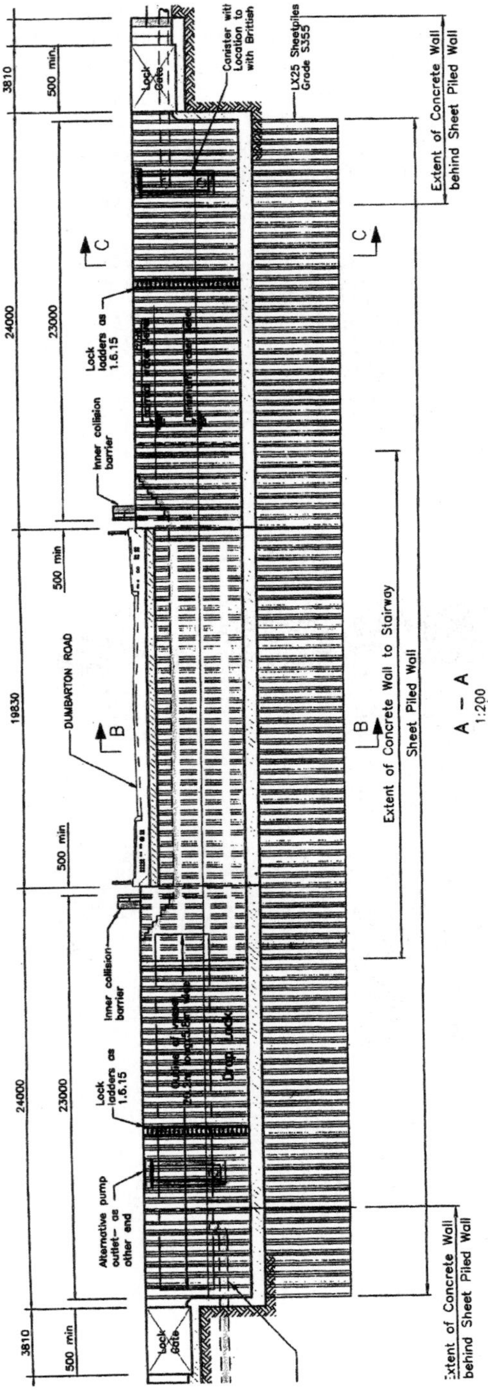
3810
500 min
Lock Gate
24000
23000
Alternative pump outlet— as other end
Lock ladders as 1.6.15
Inner collision barrier
Drag Lock
19830
500 min
DUMBARTON ROAD
B
C
Canister with Location to with British
LX25 Sheetpiles Grade S355
Extent of Concrete Wall behind Sheet Piled Wall
Extent of Concrete Wall to Stairway
Sheet Piled Wall
A — A
1:200

SKETCH 2

BONNYBRIDGE

The crossing of the canal at Bonnybridge provides the only vehicle crossing of the canal for a considerable distance in an easterly and westerly direction. The approach grade is in excess of 10% from the north and there is a T-Junction immediately to the south of the canal.

Originally it was the site of a bascule bridge and laterally a swing bridge before an ARMCO culvert closed the navigation. For a fixed crossing the road level would have to rise by 2.8m and this would greatly increase the northern approach gradient, making it impractical.

The solution proposed is to build a flat plate hydraulic lifting bridge which provides the 3m headroom requirement. This is preferred to a swing bridge because operation is faster, the span is smaller and craft transit time is less. The natural failsafe position is for the bridge to sit on its bearings in the road open position with the smaller span meaning that the T Junction to the south is able to have standard radii at the junction. Sketch 3 below shows the proposed arrangement. Because commercial craft will probably wish to operate in this section an alternative lifting height of 4.5m will be provided to allow for twin deck craft which will therefore be more economically viable.

SKETCH 3

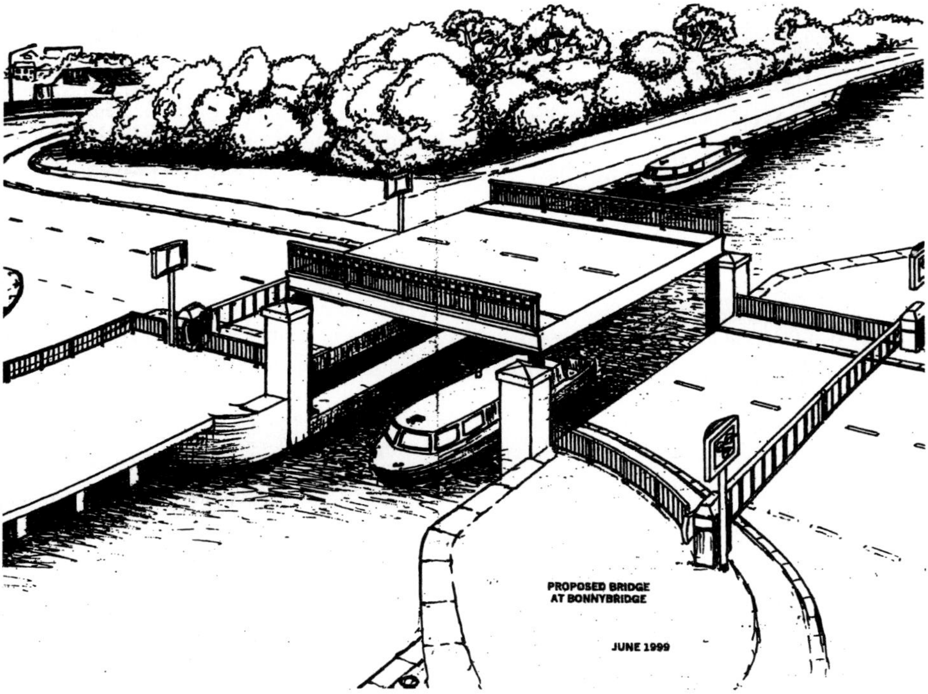

The Millennium Link: Grangemouth Link – Contract C8

LESLIE CLARKE MICE, CEng, Principal Construction Engineer, British Waterways, Technical Services, Fearn's Wharf, Neptune Street, Leeds, LS9 8PB

1. EXECUTIVE SUMMARY

This paper describes the feasibility, exemplar design and tender stages of the Grangemouth Link Project. This project forms Contract C8 of the British Waterways scheme to reopen to navigation the Forth and Clyde Canal and completes the eastern end of the work which takes the canal from just east of Falkirk into the Firth of Forth.

The approximate budget project value was some £4 million and the project encompasses a route of 6 kilometres from the existing end of the canal to the Forth adjacent to Grangemouth Docks. The feasibility study was wide ranging and the recommended option, as determined by budget and time constraints, was a route utilising the River Carron. This required a 700m length of new canal with 2 locks and a mooring basin to link the original canal to the river, followed by 5 km of tidal river navigation to enter the Forth.

The feasibility study was commissioned in March 1998 and the works contract was awarded in December 1999.

2. INTRODUCTION

British Waterways (BW) manages the major part of the inland waterway system in England, Scotland and Wales. BW's mission statement is as follows:-

"Our business is to manage the inland waterway system efficiently for the increasing benefit of the United Kingdom. We aim to provide a safe and high quality environment for customers, staff and local communities. We take a commercial approach and strive for excellence in every aspect of our work. The heritage and environment of our waterways will be conserved, improved and made to work well for future generations".

In order to operate, maintain and improve the waterway network, the area covered is split into Regions, Scotland being one of these. In order to fulfil the mission statement, BW seeks funding from as many sources as possible outside of Government funding and self generated income. At present Government funding amounts to approximately 40% of total expenditure.

The Millennium Link Project is a prime example of this approach. The project has more than 16 funders, the major one being the Millennium Commission.

The main purpose of the project is to reopen to navigation the Forth and Clyde canal, the Union Canal and the associated feeder system. The project is run from the Scotland Region office of BW by a dedicated team. It is divided up into a number of contracts; the Grangemouth Link Contract being C8.

The objective of the Grangemouth Link Project is to join the existing finish of the Forth and Clyde Canal at West Mains Industrial Estate near Falkirk with the River Carron at Grangemouth Docks and hence the Firth of Forth and the North Sea. Grangemouth is located some 12 miles west of the Forth Road Bridge as shown in Figure 1.

A requirement of the Millenium Link Project was that all contracts be carried out on a design and construct basis under the New Engineering Contract. "Exemplar" designs would be provided to tenderers to be used as a basis for progressing the design prior to construction. The initial feasibility and exemplar design stages were to be tendered for by selected

The Millennium Link. Thomas Telford, London, 2000, 86–108

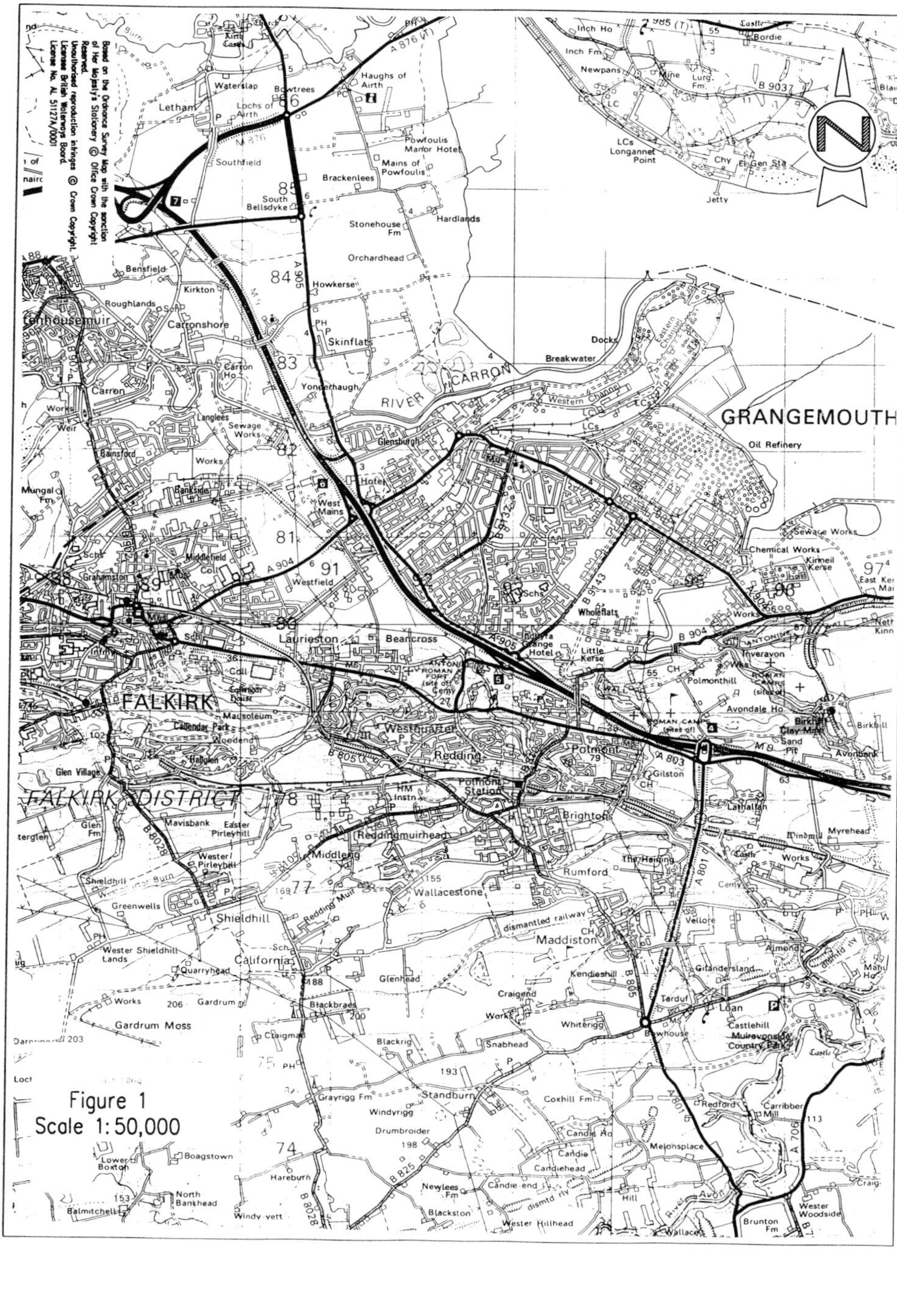

Based on the Ordnance Survey Map with the sanction of Her Majesty's Stationery © Office Crown Copyright Reserved.
Unauthorised reproduction infringes © Crown Copyright.
Licensee British Waterways Board.
License No. AL 51127A/0001

Figure 1
Scale 1:50,000

consultants. British Waterways internal engineering group based in Leeds – Technical Services – were offered the opportunity of bidding for these initial stages. After consideration of the scope and complexity of the project Technical Services decided to approach a Consulting Engineering firm in the local area with a view to offering a joint venture bid. Bullen Consultants agreed to this proposal and also offered Holford Associates (now Abbey Holford Rowe) as a third partner. After tendering, the feasibility stage was awarded to this joint group. A project team was established with the following areas of responsibility:-

BW Technical Services (Leeds)	Bullen Consultants (Glasgow)	Holford Associates (Glasgow)
• Canal and Navigation Issues • BW liaison • Mining advice • Hydrographic issues • Geotechnical issues	• Project Management • Services issues • Authority liaisons • Local issues • Tender process • Planning supervision • Surveys	• Environmental Matters • Landscaping and Architectural issues

Programme

The overall programme for the Millennium Link Project required that all construction contracts had to be awarded by December 1999and that all costruction work must be completed before April 2001.

The project programme for the Grangemouth Link section was divided into 4 main stages as follows:-

	Stage	Time Period
1.	Feasibility study, tender to completion	March 98 – October 98
2.	Exemplar Design	November 98 – April 99
3.	Tender Process	April 99 – September 99
4.	Site Works	September 99 – March 2001

Stage 1 and combined stages 2 and 3 were awarded by the client as separate packages of work and the joint venture team progressed these through to completion.

Budget

The finance allocated to the construction element of this part of the Millennium Link Project was approximately £3.8m

3. HISTORY AND BACKGROUND

3.1 History

The Forth and Clyde Canal, when fully completed, ran from Bowling west of Glasgow to Grangemouth and the River Carron on the east side of Scotland, a length of some 35 miles. The Authorisation Act of Parliament for the canal was passed in 1768 and was for vessels with a draught of 7 feet. The Engineer, John Smeaton, had drawn up plans for the canal and work on construction proceeded rapidly from east to west, with half of it completed within 2 years and 9 months. It was fully open by 28 July 1790. Subsequently it has been twice deepened to 9 feet and then to 10 feet (in 1820) and was capable of taking vessels 68ft 6" long and 19ft 8" wide. As with most canals throughout the United Kingdom it had periods of prosperity and decline, but the last important usage of the canal ended after the Second World War. The canal was closed to navigation in 1961.

Following closure of the canal it was only necessary to maintain the canal in a safe condition and to ensure water passage. Consequently sections were infilled, moving bridges were fixed in position or replaced and in places development occurred over the original canal. The plan of 1987 shows a 3 kilometre gap between the end of the canal and Grangemouth Docks.

Figures 2, 3 and 4 indicate the land usage in the Grangemouth area in 1865, 1958 and 1993 respectively and this demonstrates the extent of development in this area.

Photograph 1 showing the end of the canal at West Mains looking toward Grangemouth

3.2 Background

The land around Grangemouth is low lying and generally flat. The original canal had 1 lock between West Mains and Grangemouth and 1 lock to enter the River Carron estuary, adjacent to Old Dock.

The geology in this area consists of marine alluvia (soft clays and silts) overlying Millstone Grit and Coal Measures. In the area at the end of the existing canal at West Mains the alluvium is about 30m thick while at the River Carron/M9 motorway crossing it is about 75m thick. In the area of Grangemouth Docks it is some 50m thick. The alluvial soils generally have low cohesive strengths, medium to high compressibility and low allowable bearing pressure. There are occasional deposits of fill material over the area as a consequence of industrial development and activity and the construction of roads and railways. There are no recorded mine workings south of the River Carron between the end of the existing canal and Grangemouth. Extensive mining has occurred north of the river however.

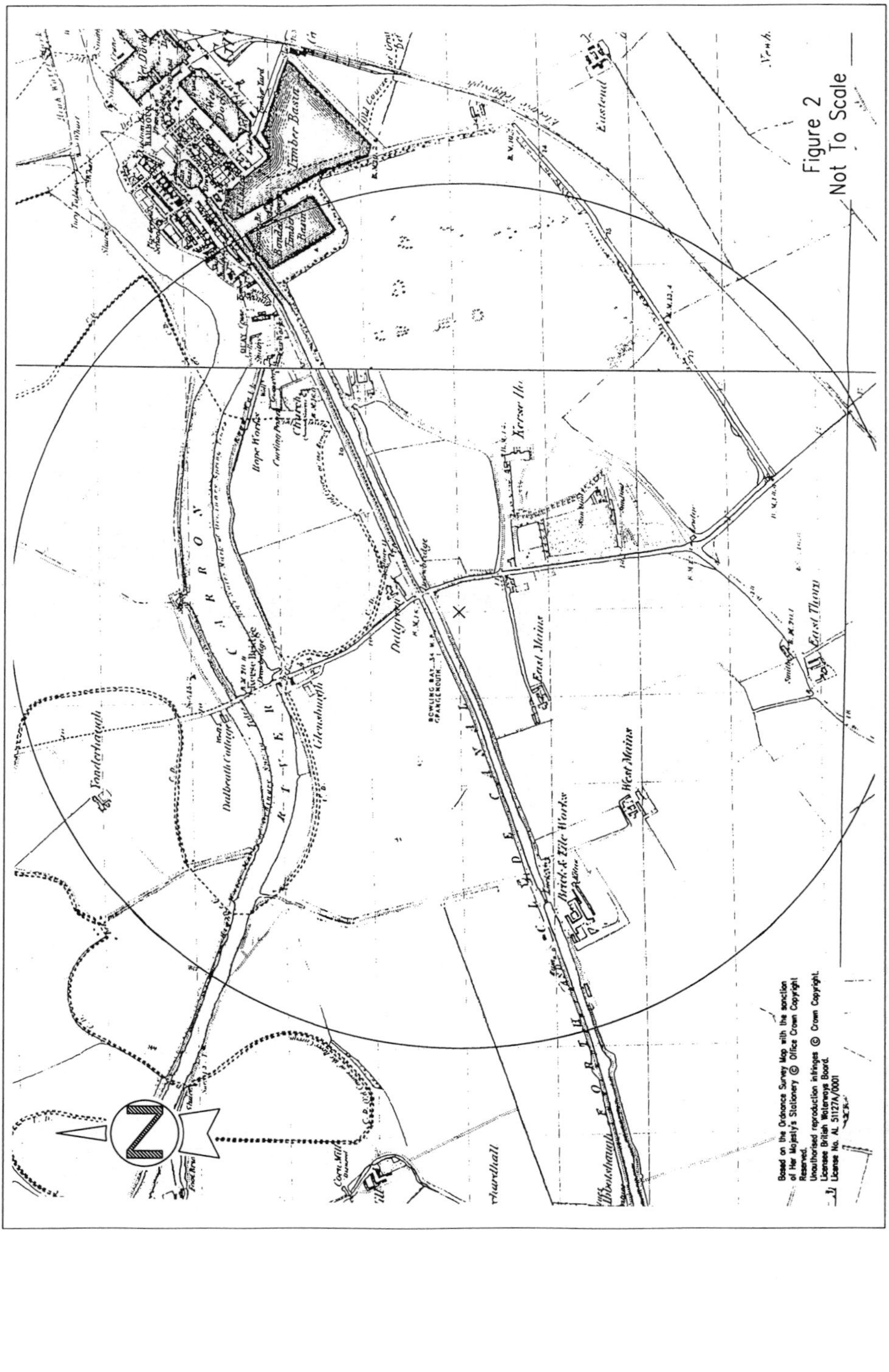

Figure 2
Not To Scale

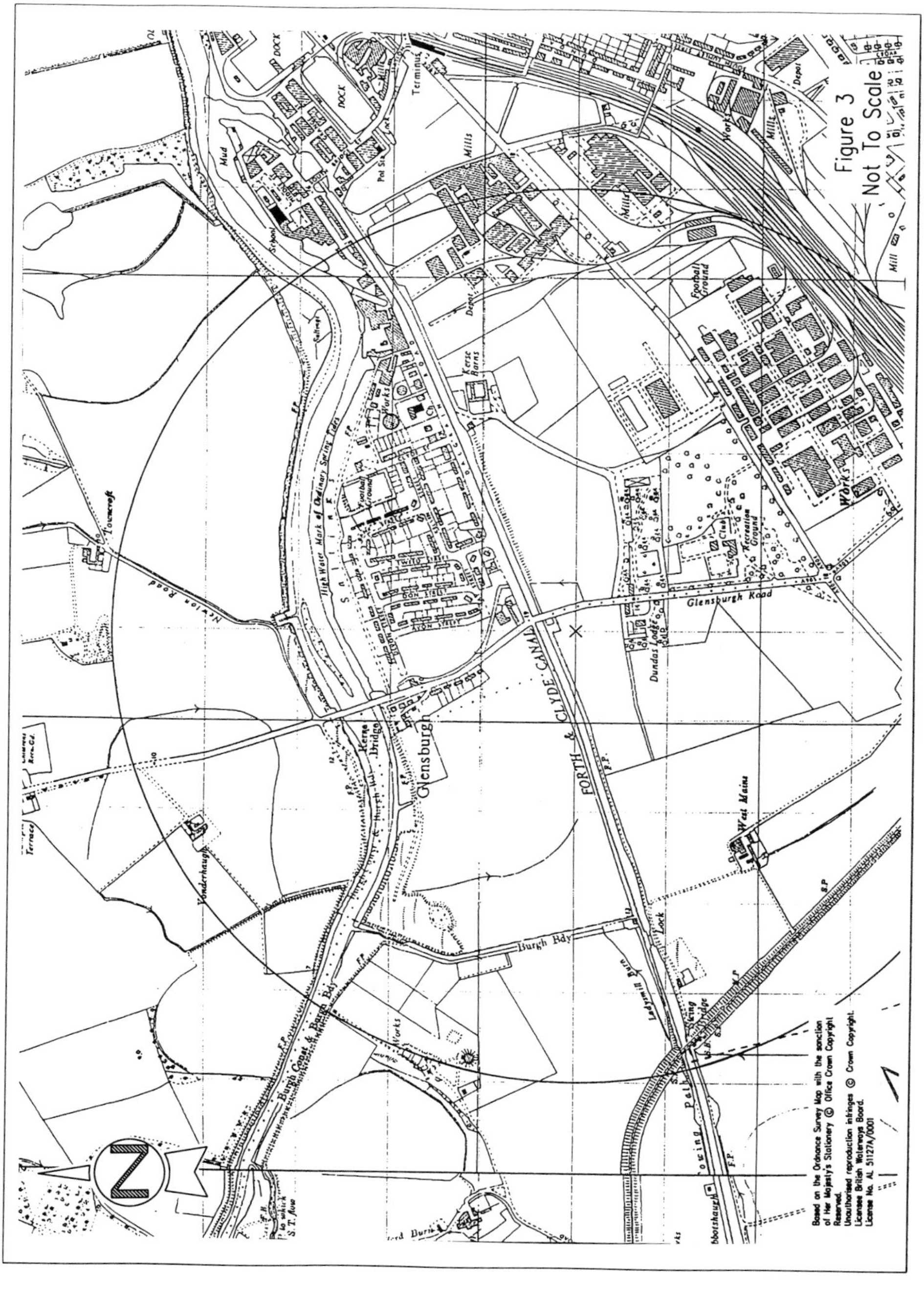
Figure 3
Not To Scale
Glensburgh
FORTH & CLYDE CANAL
Glensburgh Road
Kerse Bridge
Burgh Bdy
Burgh Coast & Burgh Bdy
Newton Road
Dundas Lodge
Recreation Ground
Football Ground
Kerse Barns
West Mains
Ladysmill Burn
Lock
Dock
Terminus
Mills
Works
Depot
School
Mud
High Water Mark of Ordinary Spring Tides
Based on the Ordnance Survey Map with the sanction of Her Majesty's Stationery © Office Crown Copyright Reserved.
Unauthorised reproduction infringes © Crown Copyright.
Licensee British Waterways Board.
License No. AL 51127A/0001

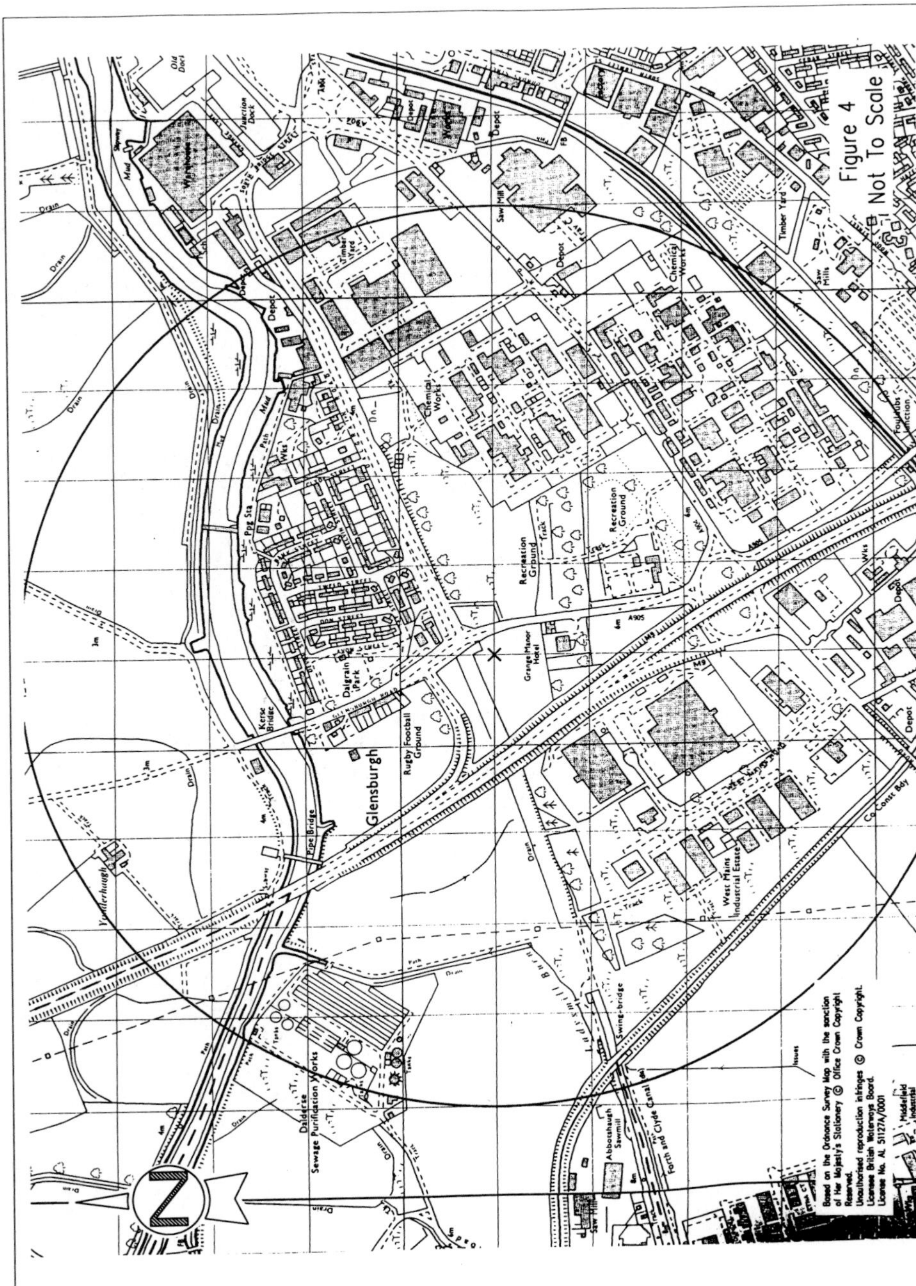
Figure 4
Not To Scale
Glensburgh
Dalgrain Park
Kerse Bridge
Pipe Bridge
Rugby Football Ground
Recreation Ground
Grange Manor Hotel
Chemical Works
Saw Mill
Timber Yard
Depot
Junction Dock
Dalderse Sewage Purification Works
Swing-bridge
Forth and Clyde Canal
Abbotshaugh Sawmill
West Mains Industrial Estate
Middlefield Industrial Estate
Yardlerhaugh
A905
M9
Based on the Ordnance Survey Map with the sanction of Her Majesty's Stationery © Office Crown Copyright Reserved.
Unauthorised reproduction infringes © Crown Copyright.
Licensee British Waterways Board.
Licence No. AL 51127A/0001

3.3 Environmental Considerations
The main environmental areas of concern are related to the River Carron estuary and the Firth of Forth. These areas are important for migratory shorebirds (waders and waterfowl) and the mudflats to the North of the Carron are designated a SSSI (Site of Special Scientific Interest). This area is also being put forward for designation as a Special Protection Area (SPA) which is a European classification for a site of the highest environmental significance.

Photograph 2 showing the SPA area to the north of the River Carron

There is also an area to the east of Grangemouth Docks at Kinneil which is an important area for waterfowl, namely shelducks. These areas are not only important for winter feeding grounds but are also important for resident bird colonies.

The River Carron is a channel for carrying treated sewage and storm water run off out into the Forth and is tidal to a point 6 kilometres inland from its confluence with the Forth. The river is bounded by tidal mud banks which have some minor value as bird feeding areas.

At a human activity level, Grangemouth has a large concentration of industrial concerns ranging from chemical works, oil refineries, docks, commercial businesses and warehousing through to residential and retail areas and all of the infrastructure associated with these.

A full environmental appraisal was undertaken for this project and this covered all aspects of the environment in the general area.

4. FEASIBILITY STUDY
The feasibility study was programmed to take place over the summer of 1998 with an option study report being presented to the client in October/November 1998. This was quite a long period of time for several reasons:-

1. the client required a thorough examination of all possible options
2. the number of organisations requiring consultation
3. the nature of the issues involved with reopening the canal

Consideration of figure 4 gives an indication of the scale of the problem in reconnecting the current end of the canal at West Mains with the River Carron at Grangemouth. The purpose of the feasibility study was to come up with costed options which could achieve this reconnection.

Three main options emerged from the study :-

1.	Use of the River Carron	Figure 5
2.	Use of the original canal route	Figure 6
3.	Use of an existing rail corridor	Figure 7

4.1 River Carron Route

This option took advantage of the proximity of the river to the canal and consisted of a new length of canal connecting to the river adjacent to the M9 motorway crossing and routed to avoid some of the existing services in the area. Two locks would be required to lock down into the river. Navigation would continue along the river to Grangemouth Docks and out to the Firth of Forth. The major constraints of this option are as follows:-

- the river is tidal and all but empties at low tides leaving little draught for navigation
- there are 2 road bridges to navigate – the M9 crossing and the Kerse Road crossing. The Kerse Road bridge has a shallow masonry invert from a previous bridge structure and at high tides there is a very small air draught for navigation
- there are 2 pipe bridges to navigate, again with small air draught at high tides

As a result of the river bed levels and the soffit levels of the bridges the river would only be navigable at certain states of the tide when there would be sufficient draught and headroom. This window of navigation was considered to be just adequate, providing information was available to boat skippers and navigation aids were used. It would be possible to improve and extend the navigation along the river by means of:-

- dredging of the river bed
- construction of a weir and associated lock at Grangemouth Docks
- construction of a barrage and associated lock at Grangemouth Docks

The last two items are undergoing further investigation and development by BW. The environmental and financial considerations concerning a barrage or weir are substantial and even on a best-case basis it would take several years to achieve a viable solution. Due to budgetary and time constraints these 2 options were excluded from the immediate project.

Investigation of the river regime revealed that under one of the road bridges the river bed was protected by a stone masonry invert, effectively forming a weir across the river during the low tide cycle. This was at a high relative level and is believed to be associated with a previous masonry bridge across the river. In order to achieve navigation this invert would have to be removed.

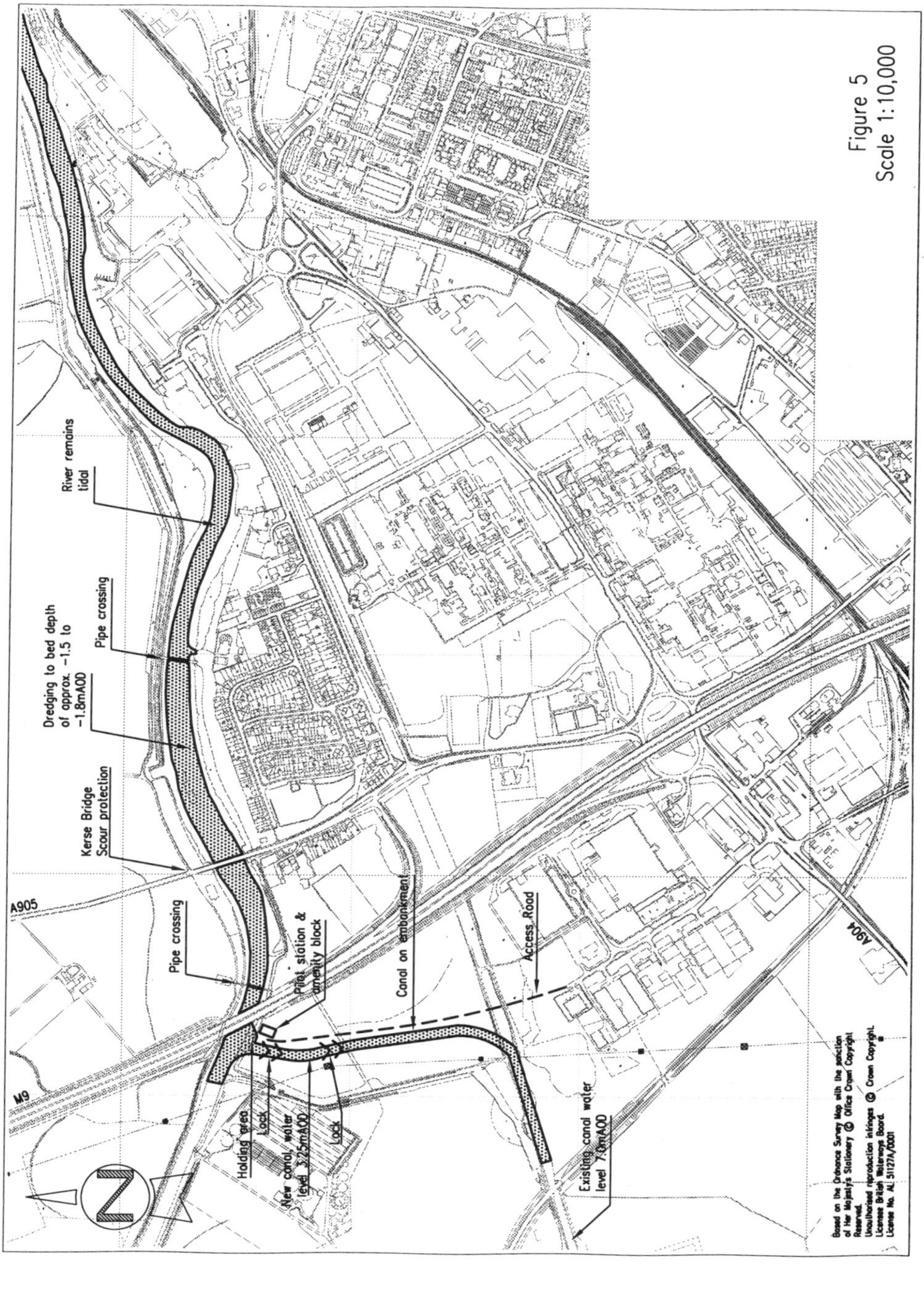

Figure 5
Scale 1:10,000

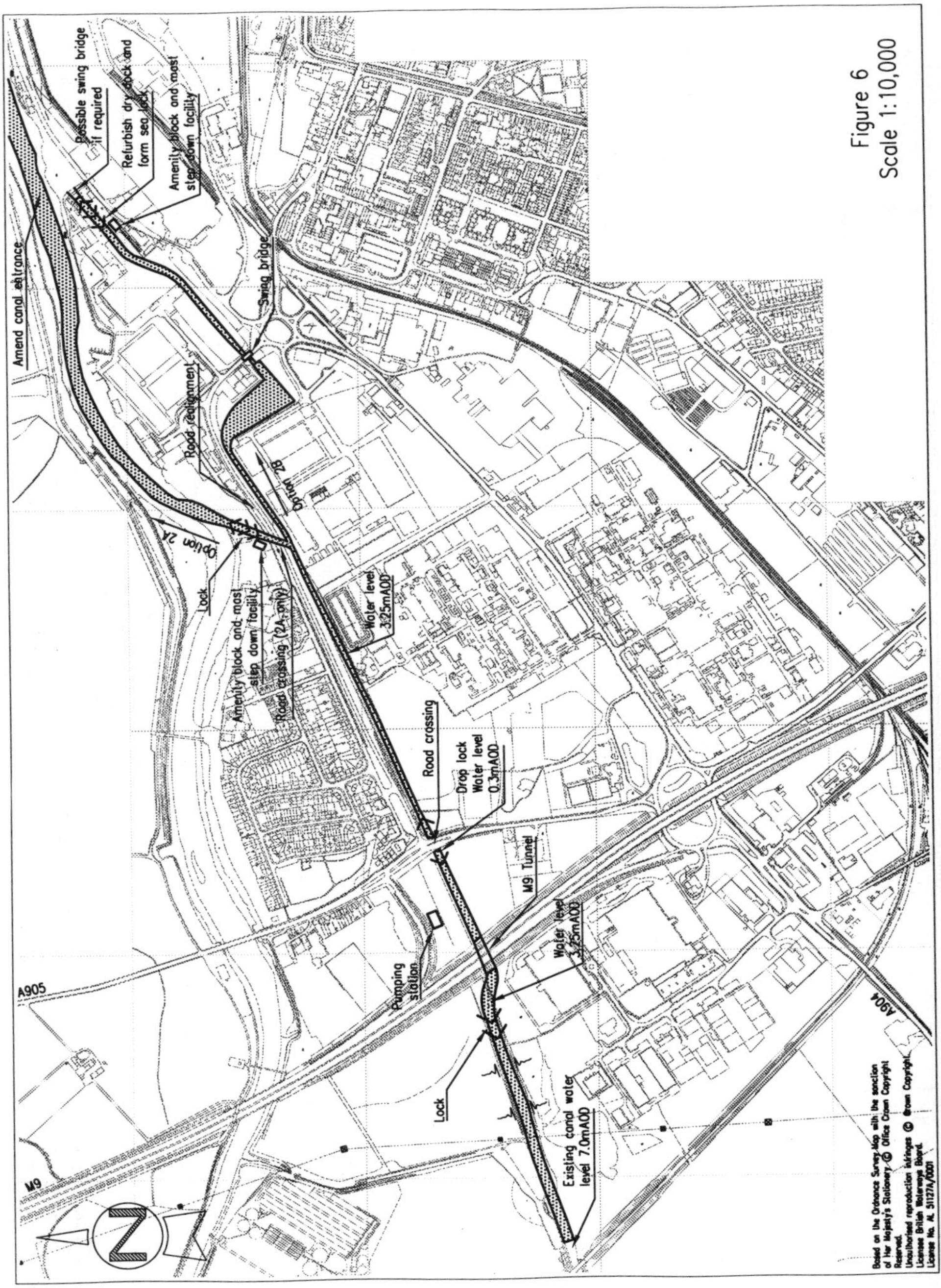

Figure 6
Scale 1:10,000

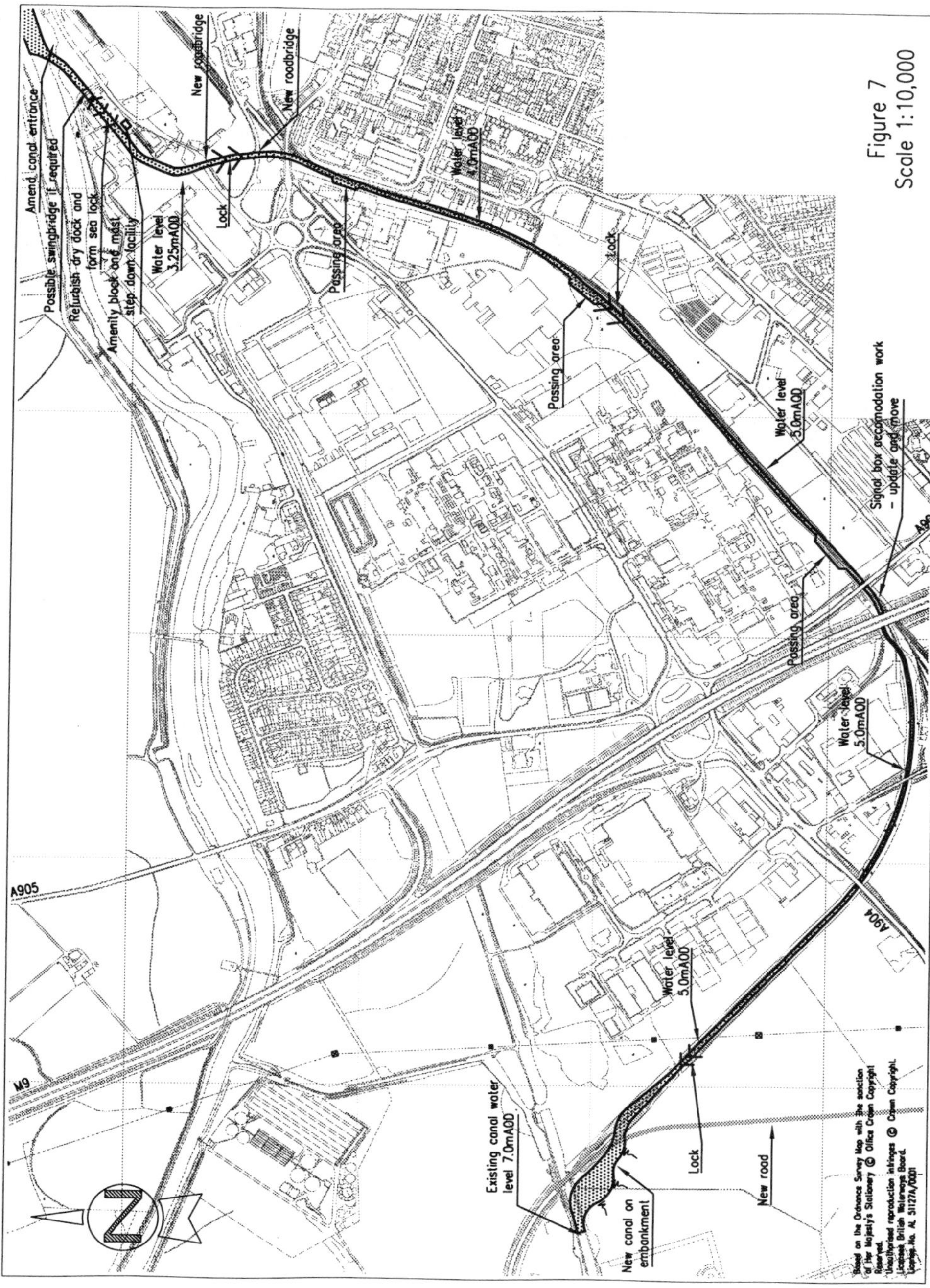

Figure 7
Scale 1:10,000

This option appeared very attractive as it made use of an existing length of water, did not involve a major amount of engineering work, involved limited disturbance to the local area and was reasonably direct.

Photograph 3 showing the Kerse Road crossing of the River Carron

4.2 Original Canal Route

This option generally follows the line of the original canal. The canal would require to be locked down to the next level to the west of the M9 motorway and pass through the M9 embankment in a tunnel. At the A905 (Kerse Road) a drop lock and pump station would be required to allow vessels to pass under the road. A new road crossing would also be required at this location.

The canal would be constructed parallel with Dalgrain Road just south of the main road, passing through a line of mature hedge and trees, with two sub-options having been considered for access to tidal waters:

- option 2A takes the canal under a new road crossing and into an old dry dock, where a new lock would exit into the River Carron.
- option 2B takes the canal further down Dalgrain Road, round the back of some unoccupied buildings and into the westernmost dock basin. A new swing bridge would be required at the entrance to the basin. The route would then proceed through the dock basins then through a new lock, built within the existing lock, thereby gaining access into the River Carron.

Photograph 4 showing Dalgrain Road, the proposed canal would be in the centre verge

There are many services which would be affected by this proposed route, including water, gas, electricity, foul sewers, surface and storm drainage, street lighting, telephone and fibre optic cables.

The route has the advantage from a historical perspective of following the general line of the original canal, but care would be required in protecting any remains of the canal, which is classified as a Scheduled Ancient Monument.

The advance works required to deal with the services would be substantial and would require a long lead-in time to resolve, not only in terms of diversions and relocations, but also in terms of agreements and management. This would have a major effect on the programme with the strong possibility that the required end date would not be met. The financial impact of dealing with the services would also be substantial.

The engineering works would be extensive and complicated by the narrow available corridor along the Dalgrain Road section. This option would present a major engineering undertaking which would generate much disruption and disturbance to road users, local residents and businesses.

4.3 Use of an Existing Rail Corridor
This option takes advantage of an existing rail corridor, which crosses the canal just west of its existing end, and passes close to the disused dock basins in Grangemouth. Three locks would be required in all, with 2 locks to facilitate passage of the canal under the A904 Falkirk road, the M9 and the A905 via existing bridge holes and then on towards Grangemouth.

Over a length of 1.5km, the canal would run parallel to a live railway line, at present used for freight traffic. Along this section the canal would be narrowed down to a minimum width with passing places.

Photograph 5. The proposed canal route would be to the left of the tracks.

In addition 2 new road bridges would be required near to Grangemouth town centre and the third lock would be situated in the Docks area itself.

As with the other options there are extensive services across the route and these would require diversions. A signal box adjacent to the M9 bridge would need to be relocated. Lead-in times would jeopardise the overall programme. This route would require fairly extensive engineering works in terms of the new canal channel, which would be just over 3km long, and the 2 new road bridges.

A considerable drawback to this option was the proposed expansion of rail freight operations in this area. This would involve bringing lines back into operation, which had been redundant for some considerable time, over land which would be required for the new canal. It appeared that any negotiations over this issue would be protracted and would have no certainty of outcome.

4.4 Option Study - Problems and Conclusions

The area around Grangemouth and the line of the original canal has altered significantly since the canal was closed to navigation in 1961. Domestic, commercial and industrial development have been steadily taking place and this has resulted in the old line of the canal disappearing, extensive services being laid and road networks being expanded and altered.

The consequence of this is that each option has significant problems to overcome, namely:-

- service diversions which would not fit in with the overall project programme
- road and service diversions would cause a lot of disturbance and disruption to local communities and businesses

- major engineering works require large financial commitments which the project budget would not be able to cater for.

All of these apply to the original canal route and the railway route options and in the final analysis ruled them out of consideration. The railway route feasibility estimate came in at almost £16 million and the original canal route was not progressed to the estimate stage due to the high probability of this option failing to meet the project deadlines and budgets.

The River Carron option had a range of estimates associated but the basic estimate came in at around £4 million. Although the estimated cost of the River Carron route was of the right order to comply with the overall Millennium Link budget, there were areas which required further consideration.
These included :-

1. How easy would the river be to navigate?
2. How extensive should the dredging be?
3. What type of dredging should be adopted in order to minimise impacts on water quality and life in and on the water?
4. What level of facilities should be provided for boats entering the canal system from the River Carron?

These questions were explored further during the exemplar design stage of the project.

The conclusion of the option study was that the only realistic solution which came close to the project budget and which complied with the time constraints was the River Carron route. The client gave approval for the choice of this option and awarded the next stage of the project – the exemplar design and tender stage- to the joint venture group.

5 EXEMPLAR DESIGN STAGE

The purposes of this stage of the project were as follows:-

1. to produce a workable design which a tendering contractor could:-
   (i) take over and progress to final design.
   (ii) adapt to suit his requirements to produce a final design.

2. to tender the project

This stage was started in November 1998 and the Tender was sent out in May 1999.

5.1 Canal Section

This route involves a new section of canal which runs from the end of the existing canal to a point just west of the River Carron M9 overbridge, and is shown on Figure 8.
The line of the canal was chosen so as to avoid :-

- high voltage overhead power lines
- existing watercourse to the west
- two oil pipelines
- the M9 bridge and embankment to the east
- a difficult junction with the river.

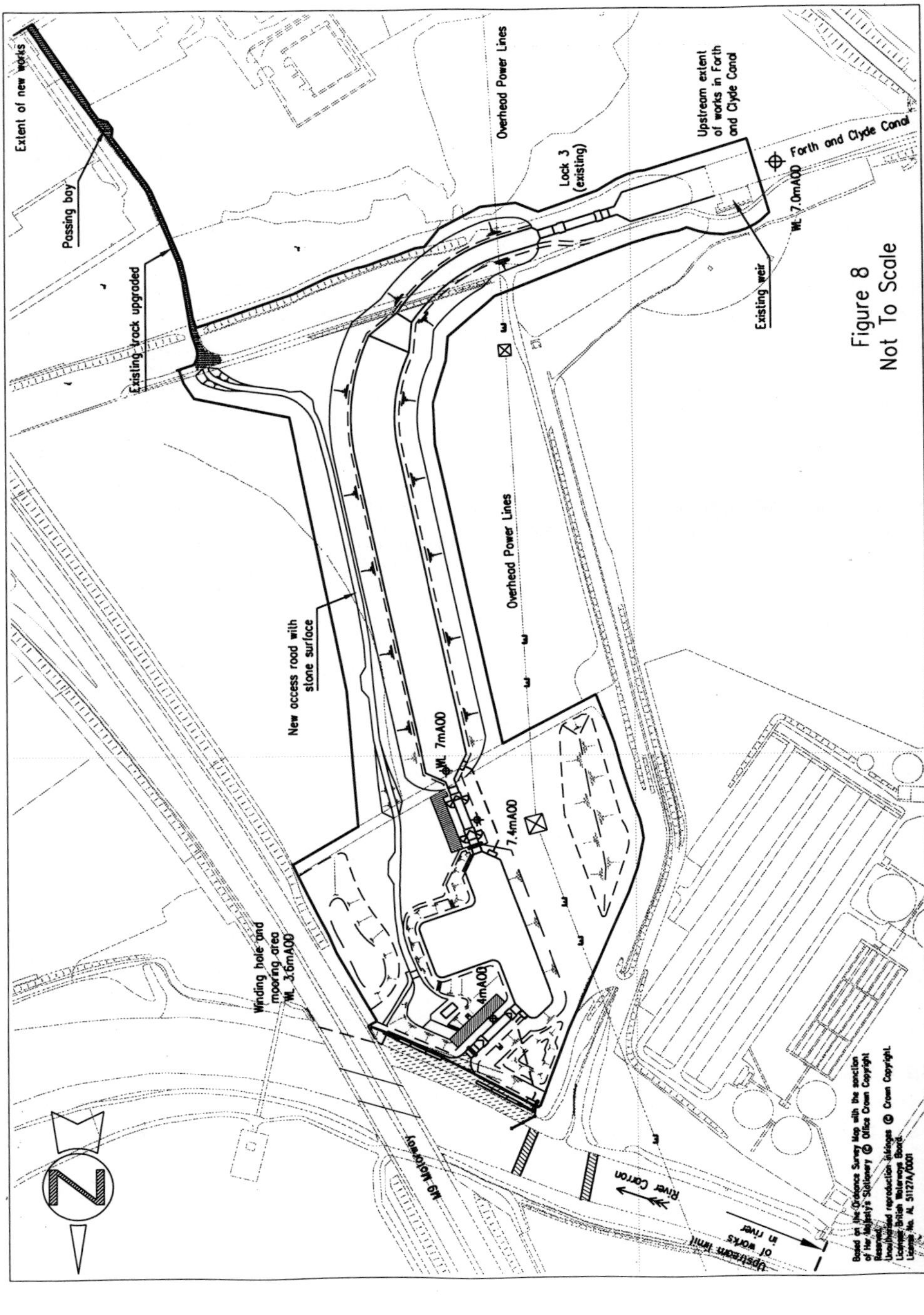

Figure 8
Not To Scale

Photograph 6. New route from the existing canal to the River Carron.

The difference in existing canal water level and river levels necessitated 2 locks along this route and these were located close together adjacent to the river. The river lock was deeper in order to accommodate the varying tidal water level in the river channel. In addition there was a requirement for an off-line mooring basin along this new canal and this was to be large enough to moor 6 full size boats. This basin was conveniently located between the 2 locks.

The new canal channel was to be mainly constructed on embankment some 4 metres above existing ground levels, utilising material excavated from the lock and basin areas in order to achieve a best use of cut and fill. The navigation channel was set at 12m width with a water depth of 2.3m. The design was a trapezoidal channel with 1 in 3 side slopes, both inside the channel and on the embankment. The preferred exemplar design included for a bituminous geomembrane liner protected by mass concrete with open-cell precast concrete blocks at water level. These blocks were chosen to promote a vegetated habitat on the edge of the new channel, a method used elsewhere in the UK very successfully.

Provision was made in the design to accommodate one of the original Forth and Clyde Locks (Number 3) some 150m east of the existing end of the canal. If salvable, this structure was to be incorporated into the new channel as a non operational lock chamber. If it was beyond renovation and use, then it would be recorded for archive and demolished. The new locks were to be reinforced concrete water retaining structures and included:-

- main chamber size: 23m long by 6.1m wide
- lock filling system
- lock by-washes
- stop plank systems
- lock ladders

- bollards
- rope slides
- lock gates
- gate quadrants

Lock 2 (inland) had the following items incorporated:-

- manually operated timber gates
- 2.8m operational water difference
- vehicular overbridge

and Lock 1 (river) the following:

- mechanically operated gates
- 4m operational water difference
- navigation aids for moving to and from the river at varying tidal states including fendering, guide piles and mooring pontoons.
- pedestrian overbridge
- lock cabin and control system

Levels were chosen at the river canal junction such that under normal flood conditions the canal would remain unaffected by floodwater and that under exceptional flood conditions the main length of new canal channel would remain unaffected, whereas lock 1 may be inundated.

Road access to the locks and the river junction was included as well as footpaths along the new canal for public access.

5.2 River Section

Moving into the River Carron, the exemplar design included for the following:-

1. a navigable channel 20m wide to a datum level of –2.3m AOD.
2. navigation markers, signage and lights down to the mouth of the river.
3. scour protection around existing bridge piers.
4. moorings, pontoons and facilities adjacent to the Old Dock at Grangemouth Docks.
5. safe anchor refuge just downstream of Kerse Road Bridge.

Photograph 7. Site of the junction of the new canal with the river; M9 and pipe bridge crossings in the background.

Various issues were associated with the dredging which was necessary for the provision of a navigable river channel.
The first issue was provision of a suitable time period for craft to navigate along the river in one pass while avoiding the problems of limited headroom under the various river bridge crossings. This was achieved by selecting a final dredged level of 2.3m below Ordnance Datum. With this scenario the client accepted that there would probably be a need for annual dredging to maintain this level, subject to the siltation regime of the river and the counterbalancing effects of tidal scour. At this stage of the project there was insufficient time to carry out a detailed model study of these siltation and scour effects.

The second issue was concerned with the historic contamination within the river sediment arising from several hundred years of industrial activity in the surrounding areas. The main compounds giving cause for concern were benzene derivatives. The conclusion of the environmental impact assessment for the project and other available information was that concentrations of these materials were low enough not to cause a problem.

The third issue related to the method of dredging. Environmental constraints required that the dredging should cause the minimum of disturbance to water quality and should not affect the SSSI areas around the estuary. The conclusion of a study into the dredging was that mechanical grab methods or similar excavation together with disposal to land or landfill would be the most appropriate.

The fourth issue concerned timing of the works. In order to minimise impact on overwintering wading bird and breeding bird populations on the shores of the Forth, dredging would preferrably be carried out only between the months of August and September.

A hydrographic survey of the river from the estuary to Dalderse sewage treatment works indicated the areas which would require removal of material to achieve the stated navigation channel dimensions. Most of this material lay between the new canal/river junction and the Old Dock area at Grangemouth. The estimate of the volume to be removed amounted to approximately 87,000m^3, based on the data obtained on the day of the survey.

Prior to the exemplar design stage it was agreed that facilities would be provided on the river at the Old Dock area at Grangemouth Docks in order to assist boats passing from sea to canal and vice versa. This was developed to include the folowing :-

- mooring area with floating pontoon and access to shore
- mooring area for demasting yachts, and mast storage area
- amenity block
- access roads and car park

6 TENDER STAGE
Tender documents were issued in the first part of May 1999 to 4 selected contractors.

The contract conditions were the second edition of the Engineering and Construction Contract (ECC), Option A, priced contract with activity schedule.

The tender process was handled by Bullen Consultants with input from the other members of the team as and when required.

Approximately halfway through the Tender stage during June, mid bid review meetings were held with the Tenderers in order to:-

1. clarify issues
2. ensure the clients expectations were being met
3. ensure the Tenderers were complying with the tender requirements
4. answer outstanding queries
5. review their design approach

Tenders were returned toward the end of July 1999 and the Tender appraisal process was started. The major concern during this process was the value of the tenders returned. The lowest offer was in excess of the allocated budget by a significant margin and there followed lengthy discussions between the parties concerned in order to try achieve increased value for money out of the scheme. This process had a finite end point and at this stage the offer values were still in excess of the budget.

Following this it was necessary for hard decisions to be taken and the client decided that various elements of the project would have to be deleted. These included the River Carron dredging and river facilities for boats. Removal of the stone invert at Kerse Road Bridge remained, as this was necessary for any navigation up river. It is anticipated that following removal of this "weir" natural scour will take place in the river and bed levels will adjust and reduce. This situation will be monitored by further hydrographic survey in order to observe the changes and to decide on any future dredging requirements. The loss of the boat facilities is not an ideal situation and the client has therefore instituted medium term plans to develop the waterway once the Millennium Link has been completed. Should finance be available, this is hoped to include the establishment of boat facilities and an enhanced navigation up the River Carron. In the longer term it may be possible to link the river navigation into the Old Dock at Grangemouth Docks and to control water levels on the river by the use of a weir and associated lock.

Following negotiation, the contract was eventually awarded in the first half of December 1999 and the target date for completion is Easter 2001, in line with the remainder of the Millennium Link Project.

LIST OF CONSULTEES AT FEASIBILITY STUDY STAGE

Utilities:-	Scottish Power Fibreway British Telecom (BT) Cable and Wireless East of Scotland Water Transco
Companies and Organisations:-	Scottish Natural Heritage Scottish Environmental Protection Agency (SEPA) Historic Scotland Falkirk District Council Scottish Office – Highway Dept Crown Estates Commissioners Forth Ports Authority Zeneca British Petroleum (BP) Railtrack B.R Property Board Alcan

MILLENNIUM LINK FUNDING ORGANISATIONS

- British Waterways
- City of Edinburgh Council
- Dunbartonshire Council
- European Regional Development Fund: East of Scotland Objective 2
- European Regional Development Fund: West of Scotland Objective 2
- Falkirk Council
- Forth Valley Enterprise
- Glasgow City Council
- Glasgow Development Agency
- Lanarkshire Development Agency
- Lothian & Edinburgh Enterprise
- The Millennium Commission
- North Lanarkshire Council
- Scottish Enterprise
- West Dunbartonshire Council
- West Lothian Council

In-the-wet construction of a new gated dam Braddock Locks and Dam, Monongahela River, PA

WILLIAM R. MILES, P.E.
Bergmann Associates; Rochester, NY
ROBERT B. BITTNER, P.E.
Ben C. Gerwick, Inc.; San Francisco, CA
WILLIAM KARAFFA, P.E.
Army Corps of Engineers; Pittsburgh, PA

INTRODUCTION

A new navigation dam will be constructed at Braddock, Pennsylvania. This project represents a significant portion of a $705 million modernization of the lower Monongahela River navigation system. The new Braddock Dam is an important component of the Corps' Civil Works program not only for the benefits it will provide to the Nation's Inland Waterways System, but also for the innovations it will bring about in the way the Corps does business. The new Braddock Dam will be constructed using innovative in-the-wet construction tech-niques to achieve cost and time savings and improve quality. The Pittsburgh District made the decision to use in-the-wet construction for the Braddock Dam in July 1997. This pioneering effort involved the use of innovative design and acquisition processes to meet the challenges in-the-wet construction would present. The Corps of Engineers and its Architect/Engineer partners successfully completed plans and specifications for the new Braddock Dam in an aggressive fast track schedule. Bergmann Associates was the lead design Architect/Engineering firm, with sub-consultants Ben C. Gerwick and D'Appolonia, employed by the Pittsburgh District for the innovative features of the Braddock Dam construction. On July 1, 1999, the Corps awarded a $107 M construction contract to the Joint Venture of the J. A. Jones Construction Company and Traylor, Bros., Inc. to construct the new Braddock Dam. This paper presents technical information on the innovative in-the-wet aspects of the Braddock Dam construction. The in-the-wet technology presented is applicable for the construction of navigation structures within existing rivers and canals around the world.

Figure 1 New Gated Dam at Braddock Locks & Dam

The Millennium Link. Thomas Telford, London, 2000, 109–116

IN-THE-WET OVERVIEW

The new Braddock Dam will be constructed utilizing two large prefabricated concrete segments that will be floated into place above, and set down upon, a pre-constructed foundation system. The foundation system is composed of sheet-pile cut-off walls and large diameter drilled shafts installed in a pre-excavated area upstream of the existing dam. This innovative in-the-wet method of construction is a technique not previously utilized for inland navigation work by the Corps of Engineers. Local and internationally recognized firms specializing in this type of construction were used by the Corps to support the government's engineering team in completing the designs, plans and specifications for this project. Traditionally, inland navigation projects have been built in-the-dry using large temporary cofferdams to dewater the area where the structure was to be built. This permitted all work to be completed within a protected dry work area and when construction was completed, the cofferdam was flooded and removed. The materials used to build the cofferdam would normally be excessed, without salvage value to the Government. Though this method is reliable, the additional costs and construction time associated with this practice prompted the Corps of Engineers to explore better ways of building these large-scale projects faster and at a lower cost to the public, and without sacrificing the quality of the end product. The Corps of Engineers conducted an exhaustive research effort to evaluate successful in-the-wet techniques used for many years for bridge piers, offshore oil and gas facilities, and immersed tube tunnels. The conclusion of this investigation was that many proven techniques, with little or no adaptation, could be used for the construction of inland waterway projects including the new Braddock Dam.

The use of innovative in-the-wet construction for the new Braddock Dam takes advantage of performing many key features of the work concurrently, rather than building the entire dam in a linear fashion within a cofferdam. This maximum use of concurrent operations will save time and thus money. The dam segments will be fabricated at an off-site casting facility in Leetsdale, Pennsylvania using a combination of precast and conventional cast-in-place concrete. Simply put, the dam will be fabricated as two hollow concrete segments that resemble large concrete pontoons. To imagine the scale of these segments, each would nearly cover a (U.S.) regulation-sized football field and will be tall enough to reach the tops of the end zone goal posts or both segments would almost cover a rugby field. When completed, these large scale floating structures will be launched and then towed to a staging pier located at Duquesne, Pennsylvania where it will be outfitted before being towed to the Braddock project site, floated into place above, and set down upon a pre-constructed foundation system.

FABRICATION OF FLOAT-IN DAM SEGMENTS

The new dam will be fabricated as two hollow concrete segments designed for flotation. These segments will be constructed off-site location while the dam and tailrace foundation systems are being completed at the project site. The float-in segments will comprise the gate sills, a portion of the stilling basin and a small portion of the pier bases. Dam Segment No. 1 will measure 101.6 m (333 ft) along the axis of the dam and will include the fixed weir bay, the water quality gate bay and one of the standard gate bays. Dam Segment No. 2 will measure 80.8 m (265 ft) along the axis of the dam and will include two standard gate bays.

Each segment will measure 31.9 m (104.5 ft) upstream to downstream and all gate bays will be 33.6m (110 ft) wide. The joint between Segments 1 and 2 will occur at Pier 3, which was made 3.4 m (11 ft) wider than adjoining piers to facilitate the connection. The float-in dam segments will be constructed with precast wall and diaphragm panels and cast-in-place bottom and top slabs. The individual wall panels of up to 6.4 m x 9.2 m (21 ft x 30 ft) and 72.6 Mg (80 tons), will be tied together with cast-in-place closure pours positioned at the intersection points of the walls.

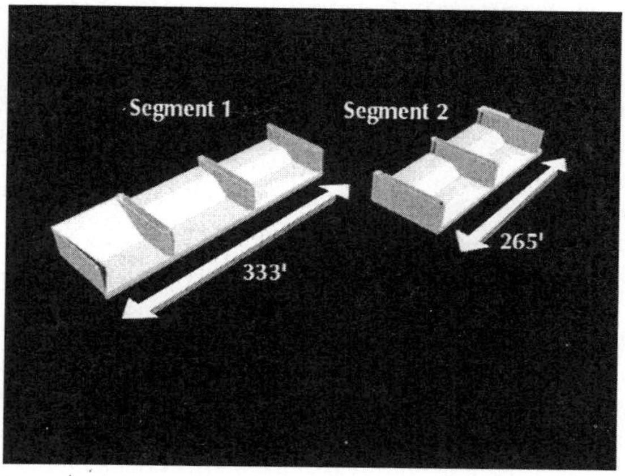

Figure 2 Braddock Float-in Dam Segments

Comprehensive evaluation of acquisition strategies to maximize the flexibility and efficiency of the construction contractor community lead to a District decision not to provide a mandated government-furnished fabrication site for the float-in concrete segments, since an array of options were possible. The contractors were allowed to determine the fabrication and assembly site and methods guided by the solicitation. Designers envisioned three possible options for assembly and launch of the float-in dam segments, including a conventional dry-dock or graving dock, a two level casting basin, and fabrication on a large offshore-type deck barge. Of the six proposals received by the Corps, four were land-based and two were barge-based.

The selected contractor elected to use the option of developing a two level casting basin. Using this method, a segment is assembled on an upper basin, a few feet above normal pool elevation in a protected area behind an earthen berm. A lower exit basin with additional depth is constructed adjacent to the upper basin. An exit structure is built at the end of the lower basin to separate the basins from the adjacent river. When the segment is ready for launch, the upper and lower basins are flooded to a common water level and the segment is floated over the lower basin. The water level in the lower basin is then returned to the outside river elevation, the exit structure is removed and the segment is towed out of the casting facility. Since the dam segments will be transported along the inland waterway navigation system, a maximum draft restriction of 3.1 m (10 ft) was stipulated in the contract documents. The Contractor will implement a rigorous program to monitor and control the weight of each segment, as they are assembled, to ensure that the maximum draft requirement is met. Once assembled, each segment will be fitted with transport bulkheads and then launched. Following a commissioning trial, each segment will be towed up river to the outfitting facility located upstream of the immediate project site. All handling and transport will be by towboats.

DAM FOUNDATIONS

While the two float-in dam segments are being fabricated off-site, work at the project site will be progressing to complete the dam foundation system. The basic foundation system is comprised of upstream and downstream cut-off walls, a graded gravel base and a grid of reinforced concrete drill shafts that extend from the riverbed into bedrock.

Eighty-eight reinforced concrete drilled shafts will carry the weight of the dam and the operating loads experienced by the dam into the bedrock. Each shaft will be 2 m (6.5 ft) in diameter and about 12.2 m (40 ft) in length. Almost 4.6 m (15 ft) of each shaft will be drilled into the bedrock to assure a secure connection and transfer of loads. Pre-excavation for the dam foundation will occur first and will consist of excavating the river bed from the existing lock river wall to the left bank abutment toe. This will also provide the necessary draft for the delivery, positioning and set-down of the float-in segments.

After pre-excavation is completed, steel sheet piling will be installed to provide both upstream and downstream cut-off walls and to serve as retaining walls for various stages of work on the dam. A prerequisite pile driving program will be used to determine the elevation of rock to which sheets are to be driven. Sheets may be ordered to accurate lengths once these elevations are determined. Using a barge mounted pile driver, steel sheets approximately 13.7 m (45 ft) in length will be installed in 10.7 m (35 ft) of water between H-piles with pre-installed interlocks at a 5.8 m (19 ft) spacing. The H-piles will be pre-driven to provide for better placement tolerance of the sheets and to shake out the sheets prior to final driving. Followers will be utilized to drive the sheets to the required top elevation without underwater cutting.

The downstream cut-off wall will then be installed in a similar manner. The downstream cut-off wall will be composed of a structural system of 0.6 m (2 ft) diameter pipe piles and sheet piles. This system is required to resist the loads imposed by the retained alluvium when the downstream face of the wall is later excavated to rock in order to install downstream scour protection. The pipe piles will be driven to rock from which point a reinforced concrete rock socket will be extended 1.8 m (6 ft) into the rock. The pipe piles will also be strategically positioned to support the downstream edge of the tailrace structure. Then a 0.3 m (1 ft) layer of 38 mm (1.5 in) crushed stone will be placed over the footprint of the foundation area. Once these operations are completed, the dam foundation system will be constructed, comprised of a drilled shaft foundation system for the dam piers and gate sills and a H-pile foundation system for the tailrace.

For the dam foundation system, two types of drilled shafts will be installed. Set-down drilled shafts will support the dead weight of the float-in dam segments at set-down and foundation drilled shafts will support the completed dam. Six set-down drilled shafts will be provided for each dam segment. Seventy seven foundation drilled shafts will be provided beneath the dam piers and gates sills. All drilled shafts will be step-tapered with a 1.8 m (6 ft) diameter rock socket drilled through a 2.0 m (6.5 ft) diameter permanent steel casing. Permanent casings will be driven and seated into the top of the claystone rock layer. Drilling will remove all material from within the casing and the 1.8 m (6 ft) diameter rock socket will then be drilled roughly 1.8 m (6 ft) into the lower siltstone rock layer. Drilled shafts will be precisely located using a pile-anchored two-level floating guide template. A 0.4 m (14 in) steel wide flange tension anchor and 0.9 m (3 ft) diameter pipe shear pin assembly which extends above the final concrete top elevation will be installed and the final concrete placement performed in-the-dry. The casing will be refilled and the top of each casing will be cut-off to its final elevation using remotely controlled cutting tools. Two rows of 0.4 m (14 in) H-piles will be driven within the footprint of the tailrace. Piles will be spaced on 2.4 m (8 ft) centers and will be driven to the claystone rock layer. The tops of these piles will extend up into the first tremie concrete pour below the tailrace slabs.

TRANSPORT OF FLOAT-IN DAM SEGMENTS

Once each dam segment is completed at the casting basin, it will be towed to the outfitting facility upstream of the Braddock dam site where the segments will be outfitted with positioning equipment and prepared for set-down onto the prepared foundations. The contractor's fabrication site is located approximately 41.9 km (26 mi) downstream of the project site. Transport time is expected to be well within the 48 hour safe period. The segments will require lockage through three locks en route to the site. Prior to actually transporting the segment, all Contractor personnel assigned to the transport operation will perform a simulated towing run using the actual tugboats that will do the towing and a barge tow made up to simulate the size of the segment. By doing this, the Contractor will be better able to understand and adjust procedures and possibly discover any unique difficulties prior to transporting the actual segments.

OUTFITTING OF FLOAT-IN DAM SEGMENTS

Once the segment arrives at the outfitting facility, it will be moored to the outfitting pier and connected to an anchor pile located approximately 152.5 m (500 ft) upstream of the outfitting facility. The anchor pile will consist of a 2.0 m (6.5 ft) diameter steel casing with a 1.8 m (6 ft) rock socket and is a contingency to protect the segment in the event of a flood while it is moored at the facility. The outfitting facility will also include a braced fendering system at least 4.6 m (15 ft) high to keep the segment on the face of the pier during high water. Dam segments will be completed and readied for set-down at the outfitting facility. Piers of each segment will be extended approximately 6.4 m (21 ft). Temporary bulkheads for immersion will be added both upstream and downstream on each gate and the fixed weir bay. Work platforms and vertical tremie pipes for post-set-down underbase and concrete infilling operations will also be installed. Once completely outfitted, each segment will draw 4.3 m (14 ft). Additional water ballast will be added as necessary to trim the segment before transport to the project site.

POSITIONING AND SET-DOWN OF DAM SEGMENTS

Once outfitted, each segment will be transported to the dam site for set-down onto the prepared foundation system. Simulated towing runs will be completed prior to transporting the actual segments. On arrival at the dam site, positioning equipment and cables will be connected to mooring piles and the dam segment will be accurately positioned over its foundations. The mooring and alignment equipment, consisting of cables and mechanical winches, will allow the contractor to make final adjustments to the alignment of each segment as they are lowered down onto their foundations. Then the hollow compartments of the segment will be gradually filled with water to add weight, causing it to sink slowly down onto its drilled shaft foundations.

After each outfitted segment is maneuvered into position, it will be ballasted onto its 6 set-down drilled shafts. A hydraulic flat jack and steel piston is located at each reaction beam. By adding water to the gate bays and using horn guides, mooring lines, snubbing towboats, and land based survey control for guidance, each segment will be accurately lowered into position. Ballast water will be fed into the dry piers until the hydraulic jacks register the required initial loading. A hydraulic ram on the downstream corner of Segment 1, adjacent to the lock wall and in conjunction with the mooring lines, will be used to align the segment on its longitudinal axis. Leveling of Segment 1 to the specified vertical tolerance will be accomplished with the hydraulic flat jack which will be operated to engage the steel piston and reaction beam. The flat jacks will be "locked-off" once the segment is fully aligned and leveled.

Figure 3 Positioning of Segment No. 1

UNDERBASE GROUTING AND INFILL OPERATIONS

Following set-down, the 0.2 m to 0.4 m (9 in to 15 in) area between the underside of each dam segment and the pre-leveled, stone-covered river bottom will be filled with grout to eliminate flow below the dam. To make the underbase grout placement more manageable, the area under segment 1 will be divided into five 21.4 m (70 ft) wide transverse strips by using inflatable bags attached to the underside of the float-in segment prior to set-down. Underbase grouting will start at the downstream row of grout pipes and continue upstream until the space below the dam is completely filled. After underbase grouting is completed, the dam segment will be "locked" onto the foundation drilled shafts by pumping a sand/cement grout into the hollow recesses surrounding the shear pin and tension anchor assemblies. Once the segments have been locked onto the foundation, the dam will be infilled with concrete which will act compositely with the hollow segment. Thirty two compartments in Segment 1 will each be filled with concrete in a two stage operation. The first stage placement consists of filling the bottom 2.4 m to 4.9 m (8 ft to 16 ft) of each compartment continuously with tremie concrete while the segment is fully flooded. The second stage placement will be placed in-the-dry after the first stage tremie concrete has cured and the compartment dewatered.

TAILRACE CONSTRUCTION

Following completion of the underbase grouting and concrete infill operations, the dam tailrace will be constructed in-the-wet utilizing precast panels that are 9.3 m (30.5 ft) wide by 6.1 m (20 ft) long by 381 mm (15 in) thick. These panels will be supported on the upstream side by the float-in dam segments and on the downstream side by the row of previously installed pipe piles, which are integral with the downstream cut-off wall. The area below the tailrace panels will be filled with tremie concrete to create a mass concrete tailrace section supported by the previously installed H-pile foundation system. Each panel will be cast with stainless steel armor angles on top surface edges and three stainless steel pipe sleeves through the panel at a 3.1 m (10 ft) spacing for tremie placement ports. To limit tremie concrete placements to 382.5 m^3 (500 yd^3) or less, a stiffened steel plate cut-off wall will be attached to the bottom of approximately every fourth tailrace panel. To provide closure to the top of downstream cut-off walls, compressible wire brushes will be connected to the downstream ends of each panel. To minimize uplift pressures on the panels during tremie placement, the tremie will be performed in two stages. The first placement will be limited to an approximate 1.5 m (5 ft) lift. Extending into this tremie placement will be a grid of rebar dowels, which will be connected to the underside of the tailrace panels. The dowels will act to provide additional support to the tailrace panels to withstand the uplift pressures caused by the second tremie concrete placement to the underside of the panels.

COMPLETION OF DAM PIERS

The upper 11.9 m (39 ft) of each dam pier will be placed in-the-dry after the float-in dam segments have been set-down and filled with concrete. This portion of the dam piers will be completed with a combination of precast concrete and conventional cast-in-place concrete. Precast panels measuring 203 mm (8 in) thick will complete the upstream portion of each pier and will function as a stay-in-place forming system. These panels will be temporarily braced in-place with sacrificial steel bracing until the infill concrete is placed. The downstream third of the pier will be conventionally formed cast-in-place concrete to facilitate the installation of the trunnion girder and cylinder girder anchorages. A pier building will be erected on the top of each pier and will house the mechanical and electrical equipment for the tainter gates and enclose the entrance into the piers interior passageways. Trunnion girders will be attached to the downstream face of each of the five piers for support of the four tainter gates. On the sides facing the gate bays, the trunnion girders will cantilever out from the face of the pier approximately 1.7 m (5.5 ft). The trunnion girders will be post-tensioned longitudinally and will also be strengthened with conventional reinforcing bars for resisting design stresses. Post-tensioned trunnion anchorage rods will extend from the girder down into the infill concrete to tie each trunnion girder to its pier.

INSTALLATION OF TAINTER GATES

Following the completion of the pier and tailrace structures and prior to installation of the left closure weir, the upstream and downstream bulkheads will be installed using a floating plant for one gate bay at a time. That bay will be completely de-watered and the tainter gate installation begun. After each bay is de-watered, a temporary steel frame will be erected to support individual tainter gate sections while they are assembled. Gate arms will be connected to their anchorages at the trunnion girders anchored to the piers . Once the assembled gate is located in its final position, two hydraulic operating cylinders will be installed and connected to each end of the tainter gate. The cylinders will also be anchored to the upper pier structure. The bay will be flooded and the bulkhead structures will be moved to another bay. All three standard gates and the one shorter and lighter water quality gate will be installed in a similar manner.

CONSTRUCTION OF LEFT CLOSURE WEIR

Following completion of the dam piers and installation of the tainter gates, the left closure weir will be constructed. This structure will connect the left end of the new dam to the existing left abutment wall providing a fixed weir crest at elevation 725. The closure weir will be comprised of two 15.9 m (52 ft) diameter sheet pile cells driven down to the top of rock. A connecting arc will be located between each cell and two closure sections between the cells, the new dam and the left abutment will complete the closure. Existing alluvium in the cells will be removed or leveled off and jet grouted and tremie concrete will be placed to the top of the sheet pile cells. After the existing dam has been breached, the lowered water level over the intervening space will allow for the construction of a variable thickness reinforced cast-in-place concrete cap to complete the weir crest.

CLOSING

This innovative construction system and its resulting benefits has applicability also for the Grand Canals of Europe. The advantages of using innovative in-the-wet construction are two-fold. Eliminating the need to build and then remove a cofferdam yield time and cost savings. Construction of the dam foundation system at the same time the dam components are being constructed offsite brings efficiency as well as cost savings to the project. As a bonus, offsite work will not be slowed by flow conditions that would normally affect in-river construction. In addition, with extensive use of high quality precasting operations, the finished dam will be more durable and have an extended useful life.

Resolving mercury contamination in the Union Canal, Scotland, UK.

N. A. SMITH, Manager, Waste Management Team, British Waterways, Leeds, UK, and
Dr O. L. LASSIERE, Environmental Scientist Scotland, British Waterways, Glasgow, UK.

INTRODUCTION

The Union Canal, which linked with the Forth and Clyde Canal and connected Falkirk to Edinburgh, provided an important transport route in the nineteenth and early part of the twentieth centuries for both goods and passengers (Lindsay, 1968). Industries which sprung up along the canal banks were able to take advantage of the transport route to both Edinburgh and Glasgow. During the height of canal side industry there were no controls for the disposal of waste. The legacy of this industrial past is a range of pollutants which are found in the canal sediments today.

This paper describes how mercury compounds were discovered by British Waterways in the Union Canal in Falkirk and the processes which have led to a resolution of this significant contamination problem under the auspices of the Millennium Link project.

HISTORY OF CONTAMINATION IN THE UNION CANAL

Dredging of accumulated sediment is an essential element of canal management to provide a sufficient depth for navigation. To aid the strategic planning of its dredging activities and to select the most appropriate routes for disposal, British Waterways conducted a national sediment survey of its 2000 mile network of canals in 1992 (Construction Industry Research and Information Association, 1996). The survey results indicated that a 9km length of the Union Canal in Falkirk was heavily contaminated with mercury. The concentration of the contamination varied along this length with the highest levels found in the central area of the stretch (Figure 1, Table 1). In all cases the total mercury concentration exceeded the Interdepartmental Committee for the Redevelopment of Contaminated Land trigger concentrations of 20mg/kg of air-dried soil for open spaces, parks and playing fields (Construction Industry Research and Information Association, 1996). This was significant as mercury is well known to be toxic to various organisms including man (Takizawa, 1979). Further sampling of the canal sediment and adjacent land confirmed the contamination problem. Analysis indicated that most of the mercury is locked up in the canal sediment in an insoluble form with very little mercury in solution (Marshall & Slowther, 1998). The spread of the mercury contamination had been limited by the obstructions to navigation at each end of the canal stretch at Greenbank Road and the A801 (Figure 1).

The Millennium Link. Thomas Telford, London, 2000, 117–124

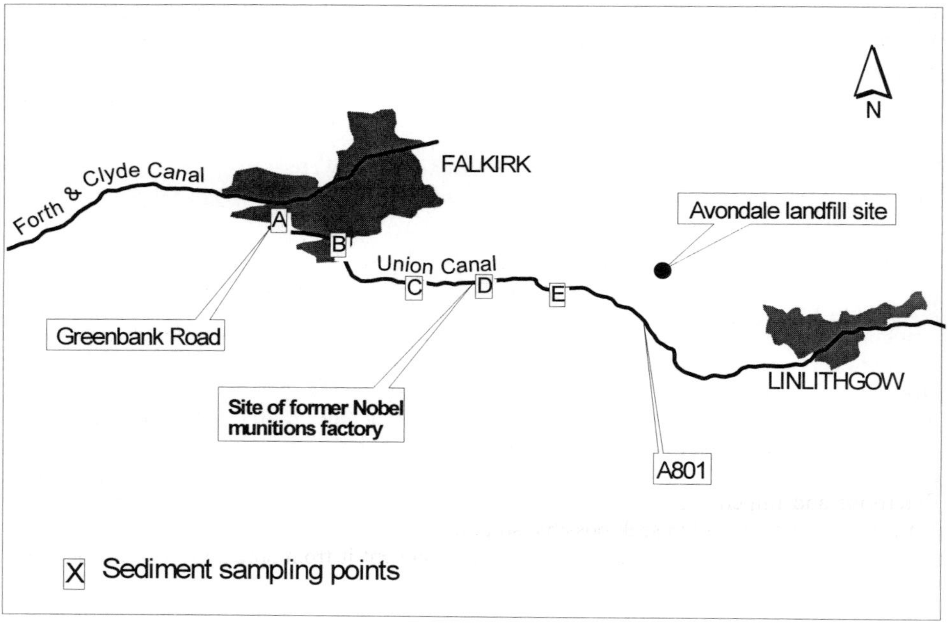

Figure 1. Map of survey sample locations showing affected area of canal

Map based upon Ordnance Survey 1:50,000 scale map data with permission from the Controller of Her Majesty's Stationery Office @ Crown Copyright. Produced by Scotland Region, British Waterways

Table 1. Range of mercury contaminants in the Union Canal

Sample location*	Total Hg mg/kg	Available Hg mg/l*	Organo Hg % of total
A	123	0.0002	0.021
B	619	0.0002	0.053
C	2360	0.0002	0.005
D	12100	0.0017	0.016
E	24	0.0002	0.146

Data from British Waterways (1994). Refer to figure 1 for sample locations.

In attempting to identify a potential source for the contamination in the canal it was established that the adjacent land in the area of highest contamination was the former site of the Nobel Westquarter factory. This factory operated between 1876 and 1968 and produced 73 million detonators, the main constituent of which was mercury fulminate (Nobel in Scotland website, April 2000). Given the local topography of the site it is likely that during the factory's life, and after its demolition, contaminated material would have had the opportunity to be transferred into the canal in solution and in a particulate form.

ENVIRONMENTAL APPROACH TO THE MILLENNIUM LINK PROJECT

When British Waterways' application to the Millennium Commission was made a major element of the bid was the environmental improvements which would be brought about by the Millennium Link project. It was recognised that sustainable use of the canal corridor would be achieved only through a full consideration of the environmental aspects of the project (British Waterways, 1995).

The award of the Millennium Link funding in 1997 provided British Waterways with the opportunity to seek a solution both to the contamination problem in Falkirk and to provide navigational depth. British Waterways' environmental policy (British Waterways, 1996) provided the basis for the approach. The internal environmental appraisal system was used to identify the issues associated with a clean-up of the canal.

IDENTIFICATION OF A SUITABLE TECHNIQUE

Following the discovery in 1992 of the mercury contamination problems, British Waterways worked with adjacent landowners and their successors, the regulatory authorities (Forth River Purification Board and latterly the Scottish Environment Protection Agency), Scottish Enterprise and Imperial Chemical Industries scientists to gain a better understanding of the scale of the problem and to seek possible solutions. British Waterways needed to find a way of dealing with the contaminated material and to prevent it from spreading any further prior to the opening of the Millennium Link .

A considerable investment was made by British Waterways to trial various techniques to remediate the contamination problem (Smith, 1998) (Table 2). Advice was sought from experts in the UK and internationally.

Table 2. Summary of decontamination trials conducted on canal sediment

Technique	Description	Category
Fractionation	Assessment of the association of contamination with particle size.	Physical
Electrokinesis	Removal of contaminants by the application of an electric current.	Physical
Soil washing	The use of water or chemicals to strip out the contaminants.	Physical/ Chemical
Volatilization*	Use of heat to release contaminants in a vapour form.	Physical/ Chemical
Cementitious stabilization	Use of cement to stabilise contaminated sediment.	Physical/ Chemical
Bioremediation with bacteria	Use of bacteria to reduce the volume of the contaminated sediment by biodegradation and to reduce the mobility of the mercury.	Biological

*this technique was considered but not trialled due to technical constraints.
Data from Smith (1998)

None of these methods proved to be practical at the full canal scale. At an early stage disposing of the material to landfill was considered but was ruled out because there were no suitable disposal sites in Scotland.

During 1999 British Waterways extensively reviewed all the available methods for dealing with the contamination. This included consultation with experts from the Department of Trade and Industry Biowise initiative for advice on methods using plants and microscopic animals. Internet searches were made for evidence of recent research and discussions were held with specialists in the field of contamination. Disposal of the contaminated sediment to landfill became a viable option at this point as a site was under construction close to the canal.

In July 1999 British Waterways held pre-tender meetings with a range of companies, who according to our research, could potentially offer a solution to the contamination problem. British Waterways did not exclude any potential contractor on the basis of their proposed technique. At this stage the proposed techniques included volatilisation, polymer treatment, lime treatment, dredging, landfill, soil washing and fractionation.

At the tender stage, potential contractors were given full details of the area of canal to be cleaned and our survey information about the contamination. In addition, a full British Waterways internal environmental appraisal was conducted which considered the potential environmental issues relating to the job. Issues which were considered included consultation with statutory bodies such as the Scottish Environment Protection Agency, Scottish Natural Heritage and local authority, informing users and neighbours, maintenance of water quality, protection of wildlife and habitats, the timing of the job, sustainability of the technique and long term effects.

On return of the tenders, British Waterways selected the contractors that offered proven technology that was environmentally acceptable and fell within the budgetary and time constraints of the Millennium Link Project. British Waterways did not have confidence in the contractors who were offering novel solutions to the problem because they had little or no experience of dealing with a situation of a similar scale.

REMOVAL OF THE MERCURY CONTAMINATED SEDIMENTS

The development of a specially engineered monocell landfill site in the vicinity of the Union Canal at Avondale, suitable to take mercury contaminated sediment led to the selection of the clean-up method. In December 1999 a contract was awarded to a joint venture made up of two specialists in their fields. These were the waste specialists Shanks Waste Services and the canal dredging experts Land and Water Services Limited. An estimated 80,000 tonnes of sediment will be removed over a 60 week period, commencing in March 2000 from the Union Canal using floating plant. The canal clean-up process is summarised in Figure 2.

The major technical difficulty associated with the waste is the liquid nature of the sediment that is not suitable for direct disposal to landfill. In this particular case, Land & Water Services Limited are using equipment which mixes wet sediment with lime (Figure 3). During mixing a chemical reaction occurs which changes the 'sloppy' sediment into a manageable soil-like form which can be transported without spillage to landfill. In addition, this form allows for compaction of material at the landfill site by heavy machinery.

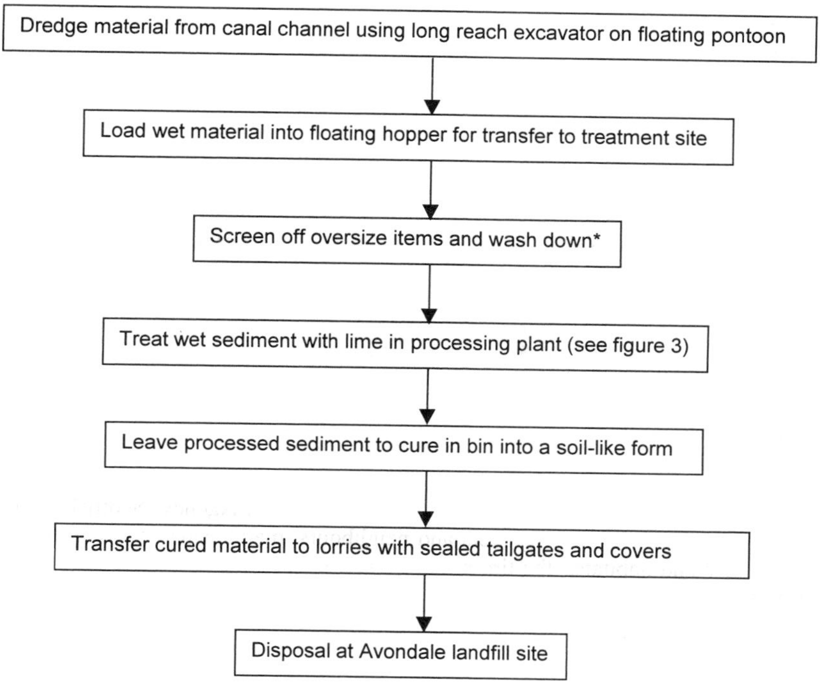

*Oversize items disposed to suitable landfill depending on level of contamination

Figure 2. Canal clean up process

As site operators, Shanks will be responsible in perpetuity to manage the waste at the site. In this location the contaminated silt will be surrounded by an impermeable layer of material and no further contamination of adjacent areas will be possible. In this exceptional waste management case, British Waterways considered it prudent to pass the responsibility for waste disposal to the experts in the field. Holding the material in a monocell has advantages in that if new technology becomes available in the future it may possible to extract and recycle the mercury in the sediment.

Spread of contaminated sediment to the adjacent clean section is controlled by the placement of a bund topped with a series of straw filled gabions to act as filters to canal water flowing to the east. The adjacent watercourses are protected from silt laden water produced during the dredging activity by sand bagging overflow weirs. During the dredging procedure, dissolved oxygen levels are recorded to ensure that they do not fall below threshold levels of 20%.

Figure 3. Lime treatment plant for silt stabilisation

Health and safety considerations
The contaminated nature of the sediment has necessitated a number of additional health and safety precautions for the contract. The risks, considered to be extremely low, are associated with handling mercury contaminated silt and the potential for the release of the toxic methyl mercury gas during the works. To protect operatives on the dredgers and around the lime processing plant special clothing is worn and the cabs of the dredger, long reach excavator and lime treatment plant have all been fitted with air conditioning filters. The treatment site is divided into clean and dirty work areas with a decontamination suite fitted with showers and changing rooms between the areas.

The contractor has produced risk assessments to ensure that the public are protected throughout the canal clean up. This will involve localised closures of the towpath and possibly some bridges. There will also be a secure fence around the lime treatment plant.

Nature conservation
British Waterways has also liaised with the government nature conservation agency, Scottish Natural Heritage (SNH), for advice about dealing with the wildlife in this contaminated stretch of canal. The vegetation is known to be contaminated and must therefore be removed to ensure complete decontamination of the canal channel. Following discussions with SNH, it was agreed that cutting back the vegetation in early spring would avoid disturbing nesting birds during the works. It was considered appropriate that fish, although contaminated,

would be allowed to swim into adjacent areas of canal (Dr Willy Duncan, SNH, personal communication). The total weight of mercury in the fish tissue is not considered to be a significant contamination problem. Water voles are also known to frequent this section of canal. The dredger operatives have been issued with identification cards to aid recognition of this species. Scottish Natural Heritage will be contacted in the event that water voles are found. When the clean up is complete, clean plants and sediment, containing seeds of aquatic plants, from areas of canal to the east will be used to aid the rapid recolonisation of the decontaminated section of canal.

Community liaison

British Waterways also recognises the importance for community liaison. Prior to commencing the decontamination contract, a community meeting was held to inform residents of the contract proposals and to provide the opportunity for British Waterways and the contractors to answer queries. Local residents have been kept informed via letter from the contractors and a number of articles have appeared in the local press and the free Millennium Link project magazine which is distributed in the Falkirk area. In addition we have been in discussion with the Seagull Trust, who offer canal cruises to disabled people, to ensure that they can continue sailing this summer during the works. Both British Waterways and Shanks Waste Services operate an open door policy and are available to answer queries about the decontamination contract at any time.

CONCLUSION

As a result of this project a major contamination problem in the United Kingdom will have been addressed. By storing the dredged material in a specially engineered landfill site the opportunity exists to recover the mercury when suitable technology becomes available.

Through a monitoring programme, British Waterways intends to measure the success of the decontamination project. The design of the new link to the Forth and Clyde Canal via the Union Canal extension and the Falkirk Wheel will limit the risk of spread of canal sediment to the Forth and Clyde Canal. The re-establishment of aquatic plants of local provenance in the cleaned channel will facilitate the recovery of the ecological integrity.

A major concern following the clean-up is the potential for recontamination of the canal from adjacent land, particularly the former Nobel Imperial Chemical Industries factory site. Redevelopment of the site and the application of new environmental legislation will present the opportunity to deal with a historical contamination problem in the Falkirk area.

REFERENCES

British Waterways, 1994. Union Canal – Redding area. A history of the investigation works undertaken to identify the extent of mercury contamination. Private and Confidential internal British Waterways Report.

British Waterways, 1995. Environment and Heritage, Millennium Link. Internal British Waterways report.

British Waterways, 1996. Environmental code of practice. British Waterways internal guidance manual.

Construction Industry Research and Information Association, 1996. Guidance on the disposal of dredged material to land. Report 157, Construction Industry Research and Information Association.

Lindsay, J. 1968. The canals of Scotland. David & Charles, Newton Abbot.

Marshall, J. & Slowther, C. 1998. Status report 1707. Assessment of sediment samples from the Union Canal. Inorganic analysis team, Imperial Chemical Industries Wilton Research Centre.

Nobel in Scotland Website, April 2000. http://www.nobel.se/alfred/dolan/index.html.

Takizawa, Y. 1979. Epidemiology of mercury poisoning. Nriagu, J.O. (Editor). The biogeochemistry of mercury in the environment. Elsevier/North Holland, 325-365.

Smith, N.A. 1998. Union Canal mercury contaminated sediments. A review of remediation options. Internal British Waterways report.

Restoration of Urban Watercourses - Watercourses in the Community – A guide to sustainable watercourse management in the urban environment.

KATHERINE A BRADSHAW, MBA, MSc, MCIWEM
Scottish Environment Protection Agency, East Region, Edinburgh, UK.

INTRODUCTION

This paper outlines the background to the launch today of a new good practice guidance booklet produced in partnership with SEPA, ICE and a number of other partners concerning the sustainable management of watercourses in the community.

The Scottish Environment Protection Agency (SEPA)

SEPA came into being on 1 April 1996 through enactment of the Environment Act (1995) which also saw the launch of the Environment Agency in England and Wales. SEPA brought together the duties of 64 predecessor organisations, which previously had regulatory responsibility for discharges to water, air, and land across Scotland.

SEPA's main aim is to provide an efficient and integrated environmental protection system for Scotland which will both improve the environment and contribute to the government's goal of sustainable development.

Whilst carrying out normal tasks associated with the licensing and environmental enforcement issues associated with these duties, SEPA staff are routinely asked to comment on development issues which may involve habitat modification. An outline survey was carried out in 1998, which established that, at the time the survey was carried out, SEPA staff were dealing with over 130 operations where an element of aquatic habitat modification was included. A summary of the results from the survey are indicated in figure 1 overleaf. The modifications noted often have serious impacts upon the landscape, amenity, fishery, wildlife and other values (for example, natural flood protection) of the aquatic habitat. The results confirm that numerous opportunities exist generally for SEPA and others to promote good practice and protect and enhance Scotland's aquatic habitats.

SEPA has responsibility, under Section 34.2 of the Environment Act 1995, generally to promote:

'*a) the conservation and enhancement of the natural beauty and amenity of inland and coastal waters and the land associated with such waters and, b) the conservation of flora and fauna, which are dependent upon an aquatic environment*'.

This duty requires SEPA staff to be aware of the wider issues associated with regulatory tasks to promote the protection of these things and wherever possible take

The Millennium Link. Thomas Telford, London, 2000, 125–130

opportunities and actions to promote environmentally sustainable practises, where such activity is not *ultra vires*.

Figure 1. Summary of SEPA 1998 Modifications to aquatic habitats survey results.

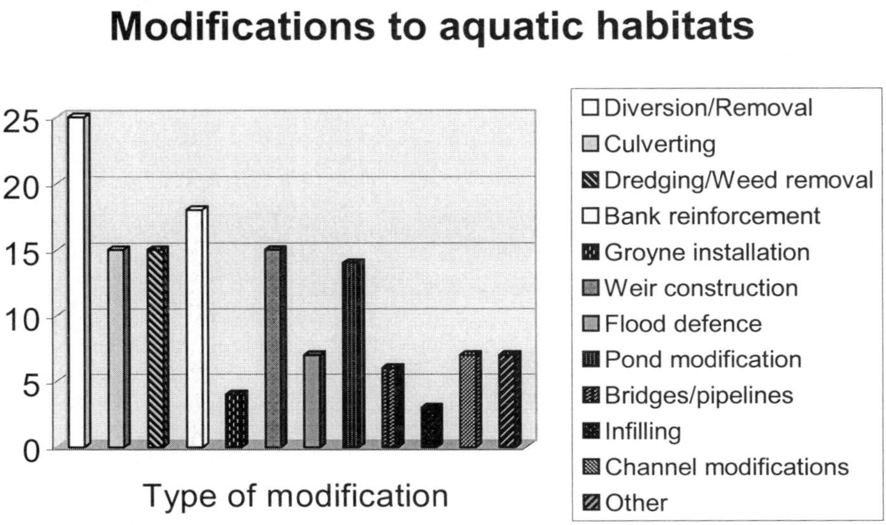

The Habitat Enhancement Initiative (HEI)
In part fulfilment of the statutory duties SEPA has under the Environment Act, the organisation launched a Habitat Enhancement Initiative in July 1998, under the directorship of W Halcrow, Director SEPA East Region. The initiative complements work already being carried out by SEPA staff, associated with the production of Local Biodiversity Action plans and has been project managed to date by the author.

Guidance produced as part of the HEI is being amalgamated into a SEPA staff handbook – the Natural Heritage Handbook, as well as being published externally. The mission statement of the HEI is indicated below:

HEI Mission Statement
'Through its own actions and in conjunction with other external partners, SEPA will advise and influence target sectors to help measurable improvement in the way in which habitats are managed, and, in particular, SEPA will focus on the aquatic and riparian habitats and the conservation and enhancement of biodiversity interests.'

The main aims of the HEI are as follows:
1. To promote conservation, enhancement and restoration of habitats in the course of SEPA's core business;
2. To build on existing SEPA projects to achieve demonstrable improvements in the management of aquatic habitats;
3. To encourage appropriate partnerships with external organisations in order to achieve improvements in the management of aquatic habitats;

4. To identify and initiate performance indicators which would allow measurement of habitat improvement targets;
5. To develop relevant internal and external guidance and support literature;
6. To train and support SEPA staff in order to complement their existing and future activities and identify and exploit opportunities for additional activities; and
7. To increase internal and external awareness though existing and new local initiatives and demonstration sites, in conjunction with other partnership organisations.

Under section 31 of the Environment Act SEPA also has the duty to promote sustainability. This factor compliments work being carried out by the local authorities within Scotland under Agenda 21. It was realised, at the start of the Habitat Enhancement initiative, that urban landscapes are increasingly unable to support a diverse aquatic ecosystem and that watercourses within these urban areas, provide an excellent opportunity both for protecting and restoring a valuable wildlife habitat. In September 1998, as part of the HEI remit, SEPA therefore commenced the production of an urban river engineering 'good practice' guidance document, with the Institution of Civil Engineers and other partners.

The main partners other than SEPA and ICE, involved with production of the document were: Dundee City Council, Edinburgh City Council, The River Restoration Centre, Scottish Natural Heritage, Scottish Wildlife Trust, The Urban Design Alliance and WWF (Scotland). SEPA is especially indebted to the Local Authorities throughout Scotland for their part in this process and for their words of encouragement received throughout the consultation process.

The conference today sees the official Scottish launch of this 'Watercourses in the Community' document, which together with this paper, pays tribute to the experience and enthusiasm of more than 80 individuals who contributed to the production process. In the words of Tom Turner, Landscape Planning, 1987:

'In the past, 'improving' rivers often meant increasing their flow capacity. In future it should refer to multi-purpose schemes designed to improve the capacity of each river valley to function as a visual amenity, a recreation area, a fishery, a nature reserve, a water supply, a storm detention area, a drainage network, and a movement corridor for boats, walkers, cyclists and equestrians'.

WATERCOURSES IN THE COMMUNITY – A GUIDE TO SUSTAINABLE WATERCOURSE MANAGEMENT IN THE URBAN ENVIRONMENT.

In order to produce a guidance document that would be credible with ecologists, planners and engineers alike, much consultation has been carried out over the last 21 months. SEPA's staff, involved with the HEI, were delighted that the Institution of Civil Engineers were prepared to head up the authorship of the document and special thanks go to Robert Huxford and Helen Taylor of ICE for their part in this process.

'Watercourses in the Community' considers both the benefits that can be obtained from the sensitive management of our urban watercourses and the harm done by neglecting them. The guidance document is a practical one, which promotes the wise use of finances as well as improving the environment in which we live. The engineering techniques advocated have been proven internationally as cost-effective, safe and practicable. The internationally recognised goal of sustainable development, to which 'Watercourses in the Community' intends to make a contribution, is mainstream policy for all political parties. Four paths to sustainability are suggested in figure 2 below.

The document outlines current techniques, philosophy and legislation regarding the sustainable management and engineering of watercourses and where to go to obtain further information and advice. The guide hopes to convince people of the importance of taking urban watercourses seriously, and to encourage them to find the advice, reference documents, guidance and further funding which will enable them to get on with the job: restoration, protection and appreciation of Scotland's urban burns and rivers.

Figure 2.

Four paths to sustainability:

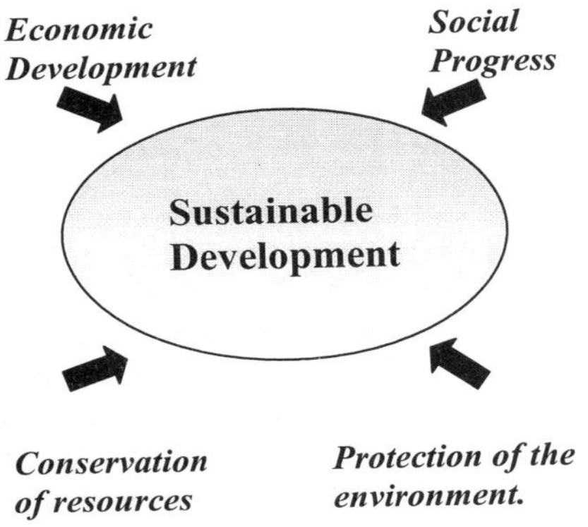

- ***Social progress*** *– by improving the attractiveness & quality of life of existing communities;*
- ***Protection of the environment*** *- by reversing the decline in natural habitats & biodiversity in urban areas.*
- ***Conservation of resources*** *– by encouraging new development in urban areas which meet peoples needs without the need for extensive car use, and provides an alternative to development on greenfield sites and in the countryside.*
- ***Economic development*** *– by increasing the profitability of development and reducing the costs of drainage and the damage caused by flooding.*

FORMAT OF THE GUIDANCE

Copies of the document have been provided to delegates today and you will note that the guidance is split into 8 main sections. These are proceeded by a knowledge map, which considers some of the key concerns relating to urban rivers and burns in Scotland, and a short introduction. The contents of the 8 main sections are as follows:

1. The importance of urban water courses – which sets out the importance of watercourses in the wider context of sustainable development and urban revitalisation, and explains why we have not made as much use of watercourses as we might;
2. Opportunities with urban watercourses – identifies the benefits from restoring urban watercourses, the practical problems that may need to be overcome, and the wide range of sustainable engineering techniques available to achieve the goal;
3. Some ideas on applying the techniques –runs through a series of different watercourse scenarios which should be of help with ideas for local schemes;
4. Working with water – covers the importance of understanding catchments and natural processes, safety, maintenance, sustainable urban drainage systems, soft engineering, and river restoration techniques;

5. Making things happen – a step by step guide to getting a watercourse restoration project going;
6. Actions – plus specific suggestions to politicians, local authorities, developers and the community;
7. Information and advice – a quick reference section to each of the subject areas, with references and lastly;
8. Glossary.

Going against the flow

It is recognised that we need to go against the flow when it comes to further degradation and habitat loss in our urban areas, if we are serious about working towards a more sustainable future. Key steps in this process include:

1. Recognising the importance of urban watercourses;
2. Establishing what people want;
3. Establishing what the problems and options are;
4. Agreeing a plan;
5. Agreeing a scheme;
6. Finding the funding;
7. Implementing it;
8. Monitoring it;
9. Managing, maintaining and enforcing it.

THE WAY AHEAD

It is however recognised that the production of a good practice brochure will never change the world, but that champions and processes need to be established in order to ensure that this good practice can be practically implemented across Scotland. This will be easier to establish if Structure, Local and Biodiversity Action Plans include the concepts being suggested, and there are good local examples available where interested parties can be directed to see the benefits for themselves. There is a need for an integrated approach to the way in which the statutory organisations address development issues, especially in light of the impending Water Framework Directive, which seeks to promote a catchment based approach for the management of the aquatic environment.

National Planning Policy Guideline 7 may emphasise that development should not proceed on river floodplains, but whilst planners continue to permit development in these areas, habitat degradation, through culverting for example, and misery to local inhabitants, caused by flooding of natural floodplains, is almost impossible to avoid.

Links with other initiatives

SEPA is active in various areas that seek to promote sustainable husbandry of the environment. Along with the organisation's policy on the restricted use of culverting, SEPA promotes a sustainable approach to the design of Urban Drainage Systems (SUDS). Scotland has led the way in this respect in promoting the use of natural techniques for draining surface water run-off from developments. As well as reducing pollution and flooding they are aesthetically pleasing in a development and can also bring about real habitat opportunities. Whilst collaboration will never be simple, it is important to recognise the need and benefits which can be achieved in doing do.

HEI Demonstration sites and Award scheme

As one means of further promoting good examples where words are put into action, the HEI has developed Demonstration Site and Award Schemes which seek to publicise good examples of sustainable aquatic management practises across Scotland. Sadly we currently have many more examples of how not to do things then we have good ones! Details of the two schemes have also been provided in leaflet form to delegates. Whilst entries were required for the award scheme by 30 June, it may still be possible to register a scheme over the next week for those involved in such work. For further information about the scheme criteria please contact Kiri Walker the HEI Assistant Project Manager on 0131 449 7296 or e-mail her on Kiri.Walker@sepa.org.uk.

CONCLUSION

It is hoped that the launch of 'Watercourses in the Community' will help mark the end to an era in which we have confined, and buried our 'liquid assets'. Perhaps it will help us move, in the spirit of partnership, one step closer towards a challenging new beginning, concerning the way in which we manage watercourses in our built-up areas.

REFERENCES

1. Habitat Enhancement Initiative, HEI Issue 1, SEPA, 2000.
2. HEI – Demonstration and Award Schemes, HEI Issue 2, 2000.
3. Protecting the Quality of our environment – Sustainable Urban Drainage - An introduction, SEPA/Environment Agency, ISBN 1-901322-12-8.
4. Sustainable Urban Drainage Systems; design manual for Scotland and Northern Ireland. CIRIA C521, ISBN 0 86017 521 9,Copyright CIRIA 2000.
5. **Watercourses in the Community – A guide to sustainable watercourse management in the urban environment. SEPA/ICE. June 2000.**

Other publications of interest:

1. Liquid Assets. Making the most of our Urban Watercourses. ISBN 07277 2792 3, 1998.
2. The Manual of River Restoration Techniques – The River Restoration Centre, Cranfield, Beds, 1999.
3. Otters and River Habitat management – 2nd edition, Environment Agency,1999.
4. Ponds, Pools and Lochans, Guidance on good practice in the management of small waterbodies in Scotland. SEPA/Pond Action, June 2000.
5. River Crossings and Migratory Fish: Design Guidance. A consultation paper, Scottish Executive, April 2000.
6. Water Vole Conservation handbook, English Nature, Environment Agency, Wildlife Conservation research Unit, 1998. ISBN 0 9529371 1 5.

Impacts of boat traffic on canal wildlife: an assessment of management options

DR. NIGEL J WILLBY,
Department of Environmental Science, University of Stirling, Scotland.

INTRODUCTION

The Scottish Lowland Canals (SLC) achieved their commercial freight heyday in the period 1820-1840. Like canals elsewhere in Britain they fell into decline after the Railway Revolution, eventually being formally abandoned as a navigation in the 1960s. Low level usage in the intervening period, combined with periodic maintenance to meet basic requirements of water supply, drainage and public safety, proved inadvertently beneficial to wildlife and the waterways became colonised by a wide range of aquatic organisms, most visibly water plants. By the 1950s naturalists were becoming increasingly aware of the nature conservation significance of little-used canals in general. At the same time, with increases in leisure time and disposable incomes, interest was growing in the revival of canals for pleasure boating. In recent decades the importance of canals as a conservation resource has grown, as comparable freshwater habitats, such as ponds and lowland rivers, have succumbed to a variety of human impacts (Willby & Eaton, 1996; Briggs, 1996). Over the same period canals have also developed into a high profile, intensively utilised, recreational resource. The Millenium Link Project showcases the potential conflicts between renewed navigation and wildlife conservation (e.g. Byfield, 1990), and the challenge of managing these conflicts during an era of heightened ecological awareness and legal responsibilities to protect biodiversity at an international level through local action. In this paper I will briefly examine impacts of boat traffic on wildlife and the conservation significance of the SLC before concentrating on possible options for managing post-restoration boat traffic impacts.

GENERAL IMPACTS OF BOAT TRAFFIC ON CANAL WILDLIFE

The process of restoration impinges on many habitats within the canal corridor, but in terms of boat traffic effects we are concerned primarily with the channel and the aquatic ecosystem. The effect of boat traffic on canal wildlife, most notably water plants, has been the subject of research at Liverpool University for over 20 years and has been reviewed on several occasions, most recently by Murphy, Willby & Eaton (1995). Canals are an ideal habitat for aquatic plant growth since they are shallow and nutrient rich, with a soft, silty bed and small, highly stable flows. Motorised boats therefore represent a high energy input to a low energy system. Because canals are small in cross section relative to the size of boats using them physical impacts on vegetation may be significant (Garrad & Hey, 1988). Chemical impacts (e.g. fuel leakage or engine emissions) may be of very local importance (e.g. close to moorings) but are insignificant by comparison.

Physical impacts take several forms ranging from: (i) direct physical damage due to contact with propellors or passing hulls; (ii) indirect effects through energy dissipated via boat wash leading to stem breakage, uprooting, or wash out of free-floating plants; (iii) high water

The Millennium Link. Thomas Telford, London, 2000, 131–138

column turbidity maintained by periodic turbulence which restricts plant growth to shallow margins; (iv) mobilisation and resettlement of fine particles produces deep, unstable, highly sorted sediments at channel margins that can inhibit aquatic plant establishment and rooting. Consequently as traffic increases there is a rapid reduction in the quantity of vegetation until by 4000 smy (standardised movments per year: annual traffic measured by lock counters, surveys or modelling, standardised to a $10m^2$ channel cross section) plants are virtually eliminated from the channel (Murphy & Eaton, 1983). Emergent plants are far more robust and substantial reed fringes can be found at the very highest traffic densities (10000 smy).

Impacts of boat traffic on species richness are rather more complex. In the total absence of disturbance from navigation or management canals experience natural infilling and species-poor reedswamp replaces open water. The rate of this change depends on factors such as the depth and the bank profile; deep vertically-sided channels, for example, tend to resist this phase for longer. In the absence of boat traffic, continued management of vegetation by dredging or weed cutting (which may form part of a long term restoration programme) will arrest succession and prevent competitive dominance by *Elodea* spp and free floating vegetation (*Lemna* spp, *Azolla*). Under such conditions aquatic plant species richness can be extremely high, often including a number of rarities (Lousley, 1976). This has sometimes given rise to the impression that boat traffic *per se* is detrimental when, in fact, a *very* light boat traffic may have similar effects on plant abundance and species richness because it substitutes a large but rare disturbance (management) for a small but more frequent disturbance that is equally effective in reducing competitive dominance.

Previous research (Pygott, 1988, Willby & Eaton, 1996) suggests that, in the absence of management, a boat traffic in the range 200-600 smy will provide a disturbance regime commensurate with high aquatic plant species richness and a density of vegetation that is not unduly restrictive of other recreational uses. Boat traffic of 1000 smy has been regarded in the past as a critical threshold for ecological damage because many aquatic plant species appear to be resistant to this traffic density, although their populations will be substantially reduced. This scenario only equates with high conservation value if species composition remains stable and there are no major declines in rare species. By 2000 smy many common species remain, but most rare species are eliminated and the habitat support function of aquatic vegetation is greatly compromised. Beyond this point, increasing traffic leads to gross simplification of the ecosystem in general and restricts the aquatic vegetation to a few highly stress tolerant species. At higher traffic canals resemble heavily engineered lowland rivers. These thresholds should not however, be interpreted over-rigorously. They are based on extensive surveys of canals subject to a wide range of boat traffic and cannot describe precisely the changes that will occur on an individual canal over time as boat traffic increases. Moreover, experience on other urban waterways suggests, for example, that steep-sided channels may be more sensitive to boat traffic impacts due to reflection and expenditure of wave energy in scouring the toe of the bank. Gently shelving rural canals with wide reed fringes that absorb wave energy may, on the other hand, prove somewhat more robust in terms of resisting species loss, even when the abundance of vegetation is greatly affected.

Our understanding of the direct impacts of boat traffic on animals and birds is rather poor and most of our knowledge derives from anecdotal observations or very localised studies. This subject is reviewed in greater detail in Murphy *et al.* (1995). It is safe to assume that the reduction in quantity and diversity of aquatic vegetation will be detrimental to the significant number of invertebrate species that rely on aquatic plants for food (directly or indirectly), shelter from predators, or as sites for egg laying and emergence. Extrapolation from other studies suggests that high turbidity generated by boat passages may lead to clogging or damage to respiratory structures, and dilution of food quality, while repeated mobilisation of the bed through propellor churning and smothering of stable substrates is likely to simplify the community to one that is typical of soft, depositional habitats. Changes in the size structure and composition of the fish community from weed-associated tench or rudd and sight feeding pike, to numerous small opportunist roach and gudgeon as traffic increases (Pygott, O'Hara & Eaton, 1990), can similarly be attributed to increased turbidity, and loss of macrophyte cover and associated large prey. Changes in habitat structure and food resource undoubtedly contribute to declines in waterbird populations but psychological stress associated with noise and visual stimuli may affect sensitive species at traffic densities below those that cause outright deterioration of habitat. Some animals, such as water voles, will persist on moderately trafficked canals (<3000 smy) provided that engineering practices are compatible with luxuriant emergent vegetation and soft banks suitable for burrowing can be retained. Higher traffic is likely to lead to bank erosion and swamping of burrows.

SCOTTISH LOWLAND CANALS: CONSERVATION IMPORTANCE

The importance of the SLC as a wildlife habitat receives full recognition in the Biodiversity Action Plans recently published or at an advanced stage of preparation, for the Regions through which the canal passes. A small section of the Forth & Clyde Canal at Dullatur is designated a Site of Special Scientific Interest (SSSI) for its aquatic plant communities, many other reaches are classified as Sites of Importance for Nature Conservation (SINC) and sections in Glasgow and Edinburgh are identified as Urban Wildlife Sites.

Aquatic plants

Victorian botanists were the twitchers of their day and took great pride in collecting and documenting the unique collection of water plants on their door step. Canals are unusual because they traverse watersheds and thereby bring into contact organisms and habitat conditions that would normally remain isolated. This provides unique opportunities for hybridisation, as evidenced by the presence of Cooper's and Bennett's pondweeds (*Potamogeton* x *cooperi* and *P.* x *bennettii*) in the SLC, their only Scottish and, in the latter case, British sites, where they have been known since the late 1800s. It also enables a remarkably high diversity of species to occur together. For example, in hosting some 12 Pondweed taxa the whole Forth & Clyde Canal qualifies technically for the status of SSSI. Canals also provide a linear corridor along which species can disperse, as can be seen from the rare tufted loosestrife (*Lysimachia thyrsiflora*) which occurs throughout most of the SLC and in Britain is virtually confined to the Scottish Central Belt. The SLC are unusual in another respect because they introduce shallow, nutrient-rich, pond-like conditions that are otherwise uncommon within the Central Belt and in Scotland generally. Hence, they form the stronghold in Scotland for some 14 species of water plant, including species like yellow water lily (*Nuphar lutea*), hornwort (*Ceratophyllum demersum*), spiked water milfoil (*Myriophyllum spicatum*), reed sweet grass (*Glyceria maxima*) and the duckweeds (*Lemna* spp) that are quite commonplace in southern England but in Scotland are rare or absent north of the SLC corridor.

Other wildlife

While the water plants of the SLC are fairly well known, information on other components of the ecosystem (which are more difficult, time consuming or labour intensive to survey properly, and whose significance maybe unclear) is very fragmentary. Sponges are a conspicuous feature of stonework, particularly around bridge holes and locks and testify to the generally high water quality. The dragonfly fauna is a similarly striking feature in the vicinity of standing waters such as Possil Loch or where the F&C abuts species rich fens near Kilsyth, and along some sections of the Union Canal and includes the locally uncommon large red damselfly (*Pyrrhosoma nymphula*). The varied composition and structure of aquatic vegetation should lead to a productive and species-rich invertebrate community and this is borne out by the existence of a healthy coarse fishery typified by perch, roach, pike and eel, and significant breeding populations of common waterbirds such as coot, little grebe, moorhen, mute swan and mallard. The greater significance of the SLC to birdlife, particularly the F&C, probably lies in the creation, together with the Kelvin Valley, of a flight path and wetland corridor, within which urban and industrial development has been comparatively restrained, and in the provision of boundary habitats attractive to more terrestrial species. In terms of mammals both Daubenton's and Pipistrelle bats use the SLC for foraging, but the water vole population on the Union canal is of particular note since this species has suffered a drastic decline in Britain as a whole over recent decades.

BOAT TRAFFIC ON THE SLC: POTENTIAL FOR CONFLICT

By applying the British Waterways-Liverpool University Boat Model network to the disposition of locks and moorings found on the SLC, and superimposing the most probable distribution of private and hire boats between mooring sites and their likely patterns of usage, it is possible to simulate boat traffic under various post-restoration scenarios. Excluding trip boats (a potentially major source of traffic on some lengths) this exercise suggests that annual average traffic will lie between 2000-3500 with a slightly lighter traffic (1500) on the Glasgow Branch. Standardising for variation in cross-sectional area (20-25m^2 on the Forth & Clyde compared to 9-11m^2 on the Union), traffic is likely to be c. 2200-2500 smy on the Union Canal, 1800 smy from Greenbank -Stockingfield Junction and 1100 and 600 smy from there to Bowling or Port Dundas respectively.

The implication, therefore, is that total boat traffic on the majority of the SLC is likely to exceed to varying degrees the thresholds previously documented for ecological damage (1000smy) and maximum aquatic plant species richness (200-600 smy). We cannot be absolutely certain how the species described in the previous section will respond to these provisional traffic densities under the specific conditions found in the SLC. However, even bearing in mind the limitations of these thresholds mentioned earlier, there is clearly potential for conflict between navigation and wildlife, particularly to the east of Stockingfield Junction. Examples of specific impacts here are likely to include a significant reduction in the quantity and species richness of aquatic vegetation, notably of floating-leaved species which are a major feature of parts of this length; loss of breeding coot and little grebe due to disturbance and decline of suitable nest sites and food; a reduction in fish biomass, notably of pike and perch with an increase in numbers of small roach. On the positive side emergent communities are unlikely to experience major adverse effects (although they will become compressed), while simulated boat traffic is sufficiently close to the ecological threshold that a range of techniques offer reasonable promise in mitigating impacts on the aquatic component of the ecosystem within selected lengths. Experience of other urban canals with comparable characteristics and vegetation composition (e.g. some of the West Midlands Canals, Coventry Canal and in Manchester the Ashton Canal) suggests that boat traffic to the west of Stockingfield Junction should be broadly compatible with sustained high aquatic plant species richness and conservation value. However, impacts on priority and potentially sensitive species should be adequately monitored.

OPTIONS TO MITIGATE BOAT TRAFFIC IMPACTS ON WILDLIFE

Limiting boat traffic

Very light traffic that is consistent with high species richness, and perhaps conservation value, appears to offer the ideal compromise between navigation and conservation interests. Unfortunately the practical relevance of this finding is to some extent lost by the fact that long-term commercial benefits of canal restoration schemes will often depend upon traffic densities exceeding those that are most desirable from a conservation perspective. The conclusion that 'some' boats are beneficial may have also served inadvertently in the past as a baseline for negotiating traffic increases beyond suitable thresholds. Currently we do not know the precise sensitivities of individual species to boat traffic under specific conditions (especially rare species). Faced with a duty to conserve wildlife generally, and specific obligations to protect and conserve certain species, we should ideally adopt a precautionary approach when recommending ecologically acceptable boat traffic densities for individual lengths. Boat traffic tends to increase and stabilise, rather than decrease over time on a given length and it is far easier to issue additional licences than revoke existing ones. Effective monitoring of traffic impacts in the early years following restoration is an integral part of this process because it allows adjustment of over-conservative limits on traffic or early detection of serious or unforeseen impacts and planning of remedial action.

At the same time, absolute restrictions on boat traffic, for example, through licence quotas or lockage restrictions, will prove controversial, and a degree of compromise is essential. As an alternative, boat traffic can be manipulated effectively so as to reduce disturbance of sensitive sites by careful control on the siting of mooring facilities and hire bases and the selective marketing and controlled development of local attractions. Improved education of boaters and the rigorous enforcement of a 4mph speed limit will both prove effective in reducing the impact of a given volume of boat traffic. Wave height and the potential for erosive wash within constrained channels increase exponentially above a critical boat velocity that is dependent on the blockage factor. Significant benefits are likely to accrue when, as in this case, traffic is close to the threshold for ecological damage. Restriction of mooring to designated sites with hardened vertical banks may be necessary to maintain the integrity of vegetation cover on soft banks and to minimise disturbance to anglers and nesting birds.

Ecologically sensitive engineering

The British Waterways Environmental Code of Practice, developed through extensive consultation with users, local managers and environmental scientists promotes good practice, integrated working and wherever possible use of habitat enhancement techniques. Such measures are not merely cosmetic and, particularly at moderate levels of boat traffic, can make a genuine difference – both ecologically and aesthetically.

Bank protection

Established fringes of emergent vegetation can dissipate 60-80% of wave energy from boat wash, depending on species composition (Bonham, 1980), and provide highly effective protection for banks against erosion, as well as nesting habitat for waterbirds, foraging habitat for water voles, and a refuge and food source for bottom dwelling invertebrates. Hence, wherever possible there should be a presumption to protect and if necessary enhance emergent plant growth, rather than replacing natural margins with steel sheet piling, which is visually intrusive and promotes bed scouring through reflection of boat wash. On heavily trafficked sections geotextile matting with pockets for planting may be needed to re-establish emergent plants on revetted soft banks, with the end product offering 'reinforced vegetative bank protection' (Brookes & Hanbury, 1990).

Dredging

Improved dredging design, recently rewarded on the Union Canal, should also in theory be beneficial to wildlife. Dredging removes deposits of soft unstable silt which are liable to be remobilised by boat wash and, by increasing the cross sectional area of the channel, should dilute the impact of an individual boat passage, particularly in terms of bed disturbance.

Bed stabilisation

Adding aggregate to the margins of heavily trafficked canals has proved highly effective in increasing the densities and species richness of invertebrate communities through the provision of a more stable and diverse substrate. The ultimate value of this technique may be to improve fishery quality by increasing growth rates of fish. This option may not be warranted at the moderate traffic densities proposed for the SLC. Simple techniques such as mowing practices that allow some cut terrestrial vegetation and attached invertebrates to fall into the canal may have significant benefits for invertebrate and fish production.

Temporal zoning

Temporal zoning is a common strategy on open waters such as reservoirs and gravel pits to reduce disturbance to wintering wildfowl by water sports. It is a non-starter on canals since the pattern of recreational use, in contrast to former commercial use, is strongly centred on the period April-September which is also the period of peak biological activity.

Spatial zoning

Overall zoning

In complex navigable systems, or large areas of open water such as gravel pits and lakes, it is possible to close off, permanently or seasonally, some areas to boat traffic without unduly restricting recreation. This option does not exist on a simple, cross-country navigation like the SLC, although the distribution of waterside tourist attractions, hire boat bases and facilities such as secure moorings results in significant spatial variation in boat traffic.

Backwaters and offline reserves

Canals incorporate various 'backwaters' such as winding holes, basins and arms, added during construction, or with the development of waterside industries, that enabled boats to pass at bridges, moor safely, manoeuvre and unload. Backwaters also include timber ponds (of which there originally several along the Forth & Clyde Canal) and sideponds designed as water storage reservoirs on lock flights, such as at Maryhill. Backwaters significantly increase the cross-sectional area of the channel and thus dilute traffic density. Most are very small (<0.1ha), but disused arms represent undisturbed mini-waterways in their own right and larger basins, as at Firhill, provide substantial areas of open water. In theory therefore, backwaters might be used to preserve examples of species-rich aquatic vegetation and associated organisms typical of lightly disturbed canals, in cases where the mainline carries high boat traffic. Backwaters may additionally act as a refuge for adult fish and a nursery area for their fry. This rationale has been used to support restoration of the Montgomery and Basingstoke canals, long renowned for the diversity of their aquatic vegetation.

The ecological value of backwaters and their potential role in canal restoration projects was assessed in detail by Willby (1994) and is discussed by Willby & Eaton (1996) and Briggs (1996). In practice the results from using canal backwaters seem likely to be generally disappointing. Backwaters serve as a focus for the deposition of silt and organic matter exported from the main channel (Bhowmik & Adams, 1989). This can inhibit submerged plant growth, or encourage encroachment of reedswamp vegetation (already adequately represented on the main channel and more resistant to boat traffic impacts) thus defeating the objective of retaining clear open water habitats. In other cases water turbidity can remain sufficiently high to restrict plant growth unless the connection with the main channel is very narrow or direct exchange of water is physically prevented. The effectiveness of backwaters for the conservation of diverse aquatic plant communities also relies greatly on a suitable management regime since intermittent or continuous light disturbance is required to arrest natural succession, firstly to dense beds of *Elodea* or *Ceratophyllum* or mats of *Lemna*, and ultimately to single-species reedswamp. However, access and management of small sites with heavy plant may prove technically difficult. Thus, Willby (1994) concluded that, without complex engineering involving silt traps and semi-permeable barriers, the potential for backwaters to make a significant contribution to nature conservation on canals was limited to larger sites within a narrow range of traffic (500-2000 smy). The potential role of disused arms as long term mooring sites for private craft was highlighted since the infrequent movement of such boats creates an intermittent disturbance regime compatible with both high plant species richness and recreational development while removing the need for interventionist management.

The disposition of existing backwaters along the SLC suggests that this option has limited potential since the most suitable sites occur in lengths where traffic impacts are likely to be modest. The lengths carrying the greatest boat traffic have very few suitable sites and there is likely to be considerable competition for these as mooring sites.

Lateral zoning
On wide channels it is possible to concentrate boats within a central track thereby reducing damage to marginal communities. In some cases this can be extended to physical protection of features, such as lily beds, using floating booms, fencing or tyre reefs (Eaton, 1986, Willby, 1994). This option could not be applied to the Union Canal but the somewhat wider Forth & Clyde Canal (14-17m) may offer limited potential within a few critical lengths.

Habitat creation
There is significant scope for imaginative habitat creation on engineered rivers as demonstrated by practical manuals such as the well known Rivers & Wildlife Handbook (RSPB *et al.*, 1994). The same scope does not exist on canals, which are frequently straight, clay-lined, elevated on embankments and lack a natural floodplain, and where there is a requirement to safeguard navigation. However, it may be possible to utilise small, low quality parcels of land, isolated between the canal and other linear features such as railways to create canal-like habitat for disturbance-sensitive species. These sites may be independent from or connected with the canal, for example through diversion of lock by-weirs. Colonisation is likely to be rapid and if necessary can be assisted by transplanting material from the main channel. On the Montgomery Canal, which is being restored to through traffic, this option has been explored in detail and found to have variable success in terms of conserving diverse open water aquatic vegetation (Briggs, 1989). Purpose-built 'offline reserves' thus provide an alternative to the use of existing backwaters, but present the same management problems of infilling and encroachment of reedswamp, and can prove costly and difficult to implement (Briggs, 1996). In political terms this approach, and to some extent that of using backwaters, tends to partition navigation and conservation functions, giving priority to navigation interests on the mainline while literally sidelining conservation issues (Briggs, 1996). An integrated approach to conservation and recreational planning from the outset is desirable but requires compromise by both lobbies.

CONCLUSIONS

The Millenium Link Project will convert the SLC into a major leisure, recreational and environmental resource. Renewed boat traffic on some lengths is likely to have adverse effects on ecosystem structure, principally through a reduction in the quantity and diversity of aquatic vegetation and increased water turbidity. In the western third of the system traffic impacts are likely to be modest but should still be closely monitored. Renewed traffic on the rest of the system will have more serious effects but, on the positive side, will be sufficiently close to previously recognised thresholds for ecological damage that several options have potential to mitigate these effects. These options include education of boaters, active enforcement of speed limits, careful siting of marinas and other facilities, general adoption of ecologically sensitive engineering techniques as embodied in the British Waterways Environmental Code of Practice, and, on some sections of the Forth & Clyde Canal, lateral zoning. Recourse to strategies such as backwaters and offline reserves, favoured in previous major restoration projects, is not advocated for practical and political reasons.

REFERENCES

Bhowmik, N.G. & Adams, J.R. (1989) Successional changes in habitat caused by sedimentation in navigation pools. *Hydrobiologia*, **176/177**: 17-27.

Bonham, A.J. (1980) *Bank protection using emergent plants against boat wash in rivers and canals.* Hydraulics Research Station, Wallingford. No. IT 206

Briggs, J.A. (1989) *British Waterways Board Montgomery Canal Ecological Survey: Reserves Report 1985-88.* British Waterways, Gloucester.

Briggs, J.A. (1996) Canals - wildlife value and restoration issues. *British Wildlife*, **7**: 365-377

Brookes, A. & Hanbury, R.G. (1990) Environmental impacts on navigable river and canal systems: a review of the British experience. *PIANC, S.I.* **4**: 91-103.

Byfield, A. (1990) The Basingstoke Canal. *British Wildlife*, **2**: 13-21.

Eaton, J.W. (1986) Waterplant ecology in landscape design. In: *Ecology & Design in Landscape*. Eds: Bradshaw, A.D., Goode, D.A. & Thorpe, E.. Blackwell, Oxford.

Garrad & Hey (1988) The effect of boat traffic on river regime. In: *International Conference on River Regime* Ed: White, R.W. John Wiley.

Lousley, J.E., (1976) *Flora of Surrey*. David & Charles.

Murphy, K.J. & J.W. Eaton, 1983. Effects of pleasure boat traffic on macrophyte growth in canals. *Journal of Applied Ecology*, **20**: 713-729.

Murphy, K.J., N.J. Willby & J.W. Eaton, (1995). Ecological Impacts and Management of Boat Traffic on Navigable Inland Waterways. In: *The Ecological Basis for River Management.* Eds: D.M. Harper & A.J.D. Ferguson. John Wiley.

Pygott, J.R. (1988) *The Effects of Boat Traffic on Canal Ecosystems*. PhD, University of Liverpool.

Pygott, J.R., O'Hara, K & Eaton, J.W. (1990). Fish community structure and management in navigated British canals. In: *Management of Freshwater Fisheries*, Eds: van Densen, W.L.T., Steinmetz, B. & Hughes, R.H.). Pudoc, Wageningen.

RSPB, NRA & RSNC (1994). *The New Rivers and Wildlife Handbook*. RSPB.

Willby, N.J. (1994) *Management of Navigable Canals for Nature Conservation and Fisheries*. PhD, University of Liverpool.

Willby, N.J. & Eaton, J.W. (1996) Backwater habitats and their role in nature conservation on navigable waterways. *Hydrobiologia*. **340**: 333-338.

The history, transition and environmental issues of the Panama Canal

DR. JERRY R. ROGERS, P.E.,
ASCE History and Heritage Committee, Civil and Environmental Engineering, University of Houston, Houston, Texas

ACKNOWLEDGEMENTS
On April 19, 1994, the Panama Canal Commission (PCC) hosted 17 members plus spouses of the ASCE Board of Direction at a briefing at the Panama Canal Administration Building. In addition, a tour of the Marine Traffic Control Center, Gaillard Cut, Gatun Lake and Dredging Division, and Miraflores Locks and operation were provided. In November 1995, Administrator Gilberto Guardia Fabrega presented an update on the Panama Canal to the American Water Resources Association Annual Conference in Houston, Texas. On October 17, 1999, Tomas W. Drohan, Director of Engineering and Industrial Services, Panama Canal Commission, presented: "Past, Present, and Future of the Panama Canal" to the ASCE National Civil Engineering Conference and Exposition, Charlotte, N.C. (extensively used in this paper). The Panama Canal Commission and recent Administrators Gilberto Guardia Fabrega and now Alberto Aleman- Zubieta (plus Tomas Drohan) are acknowledged for their engineering and communications leadership for one of the most important engineering projects in the world.

INTRODUCTION
In the U.S., two selected newspaper article headlines note: "Drawn to America's Waterways: Restored canals offer fans chance to glimpse life in America's past" and "Passed by time, canals see new life as tourist attractions." The regeneration of the Forth and Clyde and Union Canals in Scotland and U.S. canal restoration efforts are excellent examples of the large public interest and importance of preserving our historic engineering past. The Panama Canal, now operated by Panama, is a large canal project that is undergoing significant changes and improvements. This paper summarizes the history, transition, modernization program, environmental issues, and capacity expansion of the Panama Canal.

HISTORY
In his book published in 1552, a Spanish priest, Francisco Lopez de Gomara, listed Panama as a preferred route for an interoceanic canal. In the late 1800s after their success with building the sea level Suez Canal, Ferdinand de Lesseps and his French company were stalled by major problems with constructing the Panama Canal. This location had major problems including a length of 75.8 km (50 mi) and 175 m (575 ft) high mountains. Seventy-five percent of the canal workers in Panama who entered the local hospitals died of tropical diseases or other causes. Equipment and infrastructure left by the French company in 1889 were valued at $43,000,000 with $287,000,000 spent. About half of the French company excavation for the Panama Canal (30 million m^3 (1.1 billion ft^3)) or 13 percent of the total final excavation was useful to the final U.S. Panama Canal plan.

The Millennium Link. Thomas Telford, London, 2000, 139–143

In 1903, after Panama separated from Columbia with U.S. naval protection, the U.S. negotiated an arrangement with the new Panamanian government to take over the Panama Canal construction in a Panama Canal Zone. U.S. construction began in 1904 directed by the first Chief Engineer John Finlay Wallace (ASCE president in 1900). From 1905-1907, instead of the French sea level canal concept, Chief Engineer John Frank Stevens (ASCE president in 1927) selected a minority report system of locks and dams proposed by Henry Abbot to "stair-step" the Panama continental divide (input from Dr. James W. Dunn, Corps of Engineers), ("The Engineering Genius History Forgot," *ASCE NEWS*, March 2000). Stevens, an experienced railroad engineer, rebuilt the old Panama Railroad track system along the excavated Panama Canal to remove rocks and gravel. When Stevens left Panama suddenly in 1907, Colonel George W. Goethals became Chief Engineer and is credited with much of the success. From the Pacific Ocean, vessels were to enter the Miraflores Locks (moved there by Goethals upon the U.S. Navy recommendation for fear of naval bombardment), float up to the Pedro Miguel Locks, float up to and transit Gatun Lake. On the other side of Gatun Lake, vessels were to drop through the set of Gatun Locks to the Atlantic Ocean. Stevens had supported Dr. William C. Gorgas successful efforts to develop a successful system of draining ponded water, cutting back vegetation, spraying mosquito breeding areas, and screening sleeping and hospital quarters to combat yellow fever and tropical diseases. Colonel David Dubois Gaillard supervised the excavation in the Culebra Cut. After Gaillard died in 1913, the Culebra Cut was renamed the Gaillard Cut in his honor. The outstanding achievements of the Panama Canal construction include:

- Gatun Lake, the largest man-made lake in the world at the time, covered 436 km^2.
- Gatun Dam, the largest dam in the world at the time, was 32 m (105 ft) high, had a maximum width at the base of 701.2 m (2,300 ft), a length of 2.4 km (1.6 mi), and a total volume of 18 million m^3 (643 million ft^3) about half as hydraulic fill. Gatun Dam impounded the Chagras River to form Gatun Lake.
- To connect both oceans, the Panama Canal excavation totaled 178 million m^3 (6.5 billion ft^3), and about 76 million m^3 (2.7 billion ft^3) were removed from the Gaillard Cut.
- The Panama Canal locks, still among the most massive concrete structures ever built, required 3.7 million m^3 (132 million ft^3) of concrete, when concrete technology was just developing.
- The Miter Gates for the locks, which were built in place, required 40,815 Mg (45,000 T) of steel. [The French company had a late, modified canal plan for smaller lock gates, which were designed by A. Gustave Eiffel (McCullough, David, The Path Between the Seas, Simon and Schuster, 1977).] With assistance from civilian engineers Edward Schildhauer and Henry Goldmark, LTC Harry F. Hodges, Corps of Engineers, supervised the final Panama Canal gates design and construction.
- A large culvert system enables lockage operations to perform by gravity. Each transit uses 197 million l (52 million gal) of water from Gatun Lake. This large water usage will be discussed further in ENVIRONMENTAL ISSUES.

Tomas Drohan, PCC Director of Engineering, recommended engineers and engineering students read the Proceedings of the Panama Canal design and construction presented in San Francisco September 20-25, 1915 at the International Engineering Congress, organized by then Major General George W. Goethals. Drohan pointed out this major canal project impacted several of our engineering techniques in practice today.

TRANSITION

As a result of the Torrijos - Carter treaty in 1979, at noon on December 31, 1999, the government of Panama took over administration and control of the Panama Canal. To prepare for the transition, during the 1990s, most of the workers and engineers were already well-trained Panamanians. Gilberto Guardia Fabrega, a Panama civil engineer, was selected as an Administrator in 1990 and was succeeded in 1996 by Alberto Aleman- Zubieta, a Panama civil engineer, educated at Texas A&M University with degrees in civil and industrial engineering. The Organic Law of the Panama Canal Authority, unanimously approved by Panama's Legislative Assembly, regulates the Constitutional Title. These contain several provisions to allow the Panama Canal to operate as an independent entity:

- Establish an 11- member board of directors to provide policy oversight (which worked during 1998-99 with the U.S. Commission).
- Provide an independent budget authority.
- Arrange its own regulations for operation of the Canal (in consultation with labor unions) and a merit-based personnel system with a special independent regime.
- Develop independent procurement and contracting procedures that closely follow the U.S. Federal Acquisition Regulations.
- Guarantee to the existing workforce continued employment and continuity of working conditions.
- Prohibit strikes but provide a Labor Mediation Board to resolve disputes.
- Establish an independent inspector general function and an ethics system similar to the 1990's system.
- Form an international Advisory Board (11 representatives of maritime industry throughout the world) to ensure the Panama Canal management closely consults with the canal users.

MODERNIZATION PROGRAM

Due to an increase in canal traffic experienced in the last few years, the Panama Canal leaders have accelerated a $1 billion modernization program to be completed in 2002. These modernization improvements include enhancing Panama Canal capacity by:

- Rehabilitating the locks tow track. With larger and stronger locomotives needed to pull the larger and heavier ships through the Panama Canal locks, the 16,158 m (53,000 ft) tow tracks required more and stronger anchoring into the concrete base beside the locks.
- Installing a lock machinery control system (LMCS) with new Miter Gate hydraulic actuators. Replacement of the 80 existing electromechanical systems will be made with new hydraulic actuators, including the first eight located at the Pedro Miguel Locks, with a new LMCS prototype at the Control House.
- Purchasing 26 stronger towing locomotives (with eight delivered in August 1999).
- Acquiring seven stronger tug boats.
- Implementing a new state-of-the art navigation system, Enhanced Vessel Traffic Management System (EVTMS). This enhanced Global Positioning System tracking will be utilized at the Marine Traffic Operations to optimize scheduling of vessels through the Panama Canal.
- Developing a state-of-the art Hydrometeorological Forecasting Center. New models from the U.S. National Weather Service will predict Panama Canal calibrated forecasts of rainfall, streamflow, and lake levels in real time (see ENVIRONMENTAL ISSUES).
- Performing a second Gaillard Cut widening program (since the 1971 widening). In 1983, the PCC began studying the Panama Canal widening with hydraulic models built in Sweden and computer experiments using PCC pilots on an advanced ship simulator at the U.S. Merchant Marine Academy. The recommended new dimensions of the Panama Canal were: increase the width of straight segments to 192 m (630 ft), increase the width of curves of up to 222.6 m (730 ft), and increase the radii of all curves. Previously limited to one- way Panamax-size ship passage in the Gaillard Cut, the latest widening project

increases the Panama Canal capacity by 20% more ships per day. Panamax- size ships are only slightly smaller than the Panama Canal lock dimensions (33.5 m (110 ft) by 305 m (1000 ft)(which led to the use of the term: "Panamax)". The widening expands the canal capacity for 6 more vessels/day from 38 ships/day to 44 ships/day, the present lock limits. (PCC's Drohan noted the major widening involves more than four times more excavation than the Channel Tunnel between the U.K. and France.) The latest widening (totaling 33 million m^3 (1.2 billion ft^3)) is larger than the original 1904 estimate for the 12.9 km (8 mi) Gaillard Cut (then 100.6 m (330 ft) wide). Due to the use of the PCC cutter suction dredge on soft rocks, this project will be finished one year ahead of schedule. In October 1999, the Gaillard Cut Widening Program had progressed to the following percentages: dry excavation was 91% complete, drilling and blasting was 86% finished, and dredging was 65% complete. The world's largest dipper dredge, Christensen, is scheduled to complete its work in Dec. 2001, 12 months ahead of schedule.

ENVIRONMENTAL ISSUES

A wide range of environmental issues and concerns exist for the Panama Canal. The Panama Canal Authority will carefully dispose of the excavated material from the canal widening program with reforestation of the impacted areas. The new Panama Canal Authority must oversee the administration, use and conservation of the water resources of the Panama Canal watershed. The Panama Canal Authority will control activities along the banks of the Canal to ensure compatibility with Canal operations. Panama Canal vessel spills and countermeasure plans are developed along with lock fire-fighting capabilities. Environmental education (for efficient use and sustainability of resources and the environment) for Panama residents is also needed. The Panama Canal is implementing a new Hydrometeorological Forecasting Center for better estimates of rainfall, streamflow, and lake levels. There is an eight month wet season followed by a four month dry season in Panama. The Panama Canal leaders are concerned over past floods and droughts, with severe droughts limiting the amount of lockage water from Gatun Lake (see INCREASING CANAL CAPACITY).

One March 5, 2000 newspaper article noted: "Panama aggressively pursuing ecotourists." A new $25 million Gamboa Rainforest Resort (with an ecology library) (www.gamboaresort.com) is being developed in the Soberania National Park, along the Chagres River. Nature friendly vacations with adventure tours and educational programs will focus on Panama's flora and fauna. The Smithsonian Institution will manage a tropical research institute at the resort village. A completed $2 million, 0.56 km (3/8 mi) aerial tram at Gamboa will take guests on a guided animal and plant tour through the jungle. A marina will offer fishing trips and educational marine life tours. A former U.S. radar tower has been converted to a seven-room hotel, the Panama Canal Rainforest Canopy Tower, for ecotourists (www.canopytower.com). The Audubon Society has named the Panama Canal Canopy Tower one of the 10 best places in the world for bird watching. Officials also plan an ecotourism resort on Juan Gallegos island in Gatun Lake. For more information, contact the Instituto Panameno de Turismo: www.panamatours.com).

Finally, from the Gatun Lake supply, the Panama Canal Authority is evaluating alternatives to the routine use of 197 million l (52,000,000 million gal) per transit. One water conservation possibility is to build a set of synchrolift locks (which conserve water) for smaller vessels than the large Panamax ships. This last point leads to a final category of INCREASING CANAL CAPACITY.

INCREASING CANAL CAPACITY

There are several alternatives to increase the Panama Canal capacity:

- Build new locks with larger chambers to accommodate ships larger than Panamax size.
- Build new locks with Panamax- size chambers.
- Build new locks with chambers smaller than Panamax size.
- Build a set of synchrolift locks (which conserve water) for smaller vessels than the Panamax size (in ENVIRONMENTAL ISSUES).

Any of the new locks alternatives (with a third set of new locks near the existing two sets of locks) will require additional water. An article in the March 2000 N.S.P.E.- U.S. issue of ENGINEERING TIMES: "Panama Canal Society leaves legacy of engineering feats" indicated one alternative plan involves building two new man-made lakes to increase the Panama Canal water storage and to produce more hydroelectric power. During the April 1994 ASCE Board of Directors tour of the Panama Canal Administration Building, the PCC had a photograph showing an excavated area about 152.4 m (500 ft) from the existing canal locks where a set of third locks were started, but halted by WWII. For more information, check the PCC webpage www.pancanal.com/ with detailed historical information.

SUMMARY

In this paper, the regeneration of two important canals in Scotland is briefly linked to the history and modernization of the Panama Canal. The first topic summarized the early history of the Panama Canal. The smooth transition (of the Panama Canal from U.S. to Panama control) was outlined with the modernization program. Environmental issues were listed with recent developments. Increases in the Panama Canal capacity with the Third Locks Projects are being considered. The Panama Canal Authority is working to preserve and improve a historic international trade canal.

Flash locks to millennium wheels

Or overcoming canal grade changes – an international comparison

DR. A. L. PRASUHN
Professor and Head of Civil Engineering, Lawrence Technological University, Southfield, MI, USA.

INTRODUCTION AND EARLY HISTORY

Canals for irrigation or water supply, and to a lessor degree navigation, represent one of the earliest areas of civil engineering. The first canal is undoubtedly lost in antiquity. The earliest known canal is often associated with King Menes and the founding of Memphis c. 2925 BC. The canal is allegedly part of a water supply system for Memphis along with a 15 m (49 ft) dam across the Nile. There were many early canals in the vicinity of Mesopotamia; the most famous of which was probably the Nahrwan Canal between the Tigris and Euphrates rivers. This canal was constructed c. 2400-2200 BC with a length of 300 km (185 miles) and a breadth of 122 m (400 ft). [7,17]

In the 7th century BC an 80 km (50 mi), 20 m (66 ft) wide, stone-lined canal was constructed by Sennacherib to transport water from Bavian to Nineveh. During the 5th century BC, Darius I completed a canal between the Nile and the Red Sea that was probably begun as early as the 7th century. [7]

Impressive Chinese canals were constructed beginning about the 3rd century BC to tie the inland regions to the major rivers and to each other. Again, there is much that is not known about these canals. The Ling Ch'u Canal was constructed about 215 BC to connect the Hsiang and Li rivers as part of an all-water route from Hunan on the Yangtze River to Canton. The original function was to transport military supplies. The Han Canal in Honan (constructed 206 BC - 220 AD) and known in later times as Old Pien Canal was another early example. However, the most famous of these canals is the Grand Canal of China. Although construction began as early as the 4th or 3rd century BC, a length of 966 km (600 mi) was not in place until 610 AD. This canal is now part of a waterway system extending for 1747 km (1085 mi). [7]

For military, as well as other purposes, the Romans constructed canals and made river improvements in England, France, Italy, and the Netherlands. First century examples include a 37 km (23 mi) canal between the Rhine and Meuse constructed by General Corbulo to avoid using the stormy North Sea, and a canal designed by Roman Counsul Marcus Drusus between the Rhine and Yssel intended to reduce flooding on the Rhine. Of course, in England the Romans connected the River Cam with the Ouse with a 13 km (8 mi) long canal, the Nene with the Witham by an 40 km (25 mi) canal, and the Witham with the Trent by the still functioning Fosse Dyke. [7]

The Millennium Link. Thomas Telford, London, 2000, 144–151

The primary similarity of all the foregoing canals, at lest as originally constructed, was a near absence of elevation changes other than those associated with the grade of the canal. If this gradient were too great, single gates similar to the flash lock described below were placed at regular intervals. They were all essentially contour canals, occasionally making use of aqueducts when crossings were required. It is likely that early river and/or canal systems at different elevations occasionally occurred in proximity. Goods would have been transhipped or, when possible, the boats raised or lowered by towing them over the barrier manually or with windlasses. Hence, canal inclines date back to antiquity as well.

As canals expanded in many areas of the world, it was essential that elevation changes be incorporated directly into the canal system. The most common solution, the traditional pound lock, will be discussed shortly. However, it is appropriate to first mention the flash lock. After the Roman period, canal building largely disappeared, but revived during the 12th century as commerce expanded. It is likely that some form of flash locks appeared in Europe at this time to improve river navigation.

FLASH LOCK

The standard pound lock is the classical as well as by far the most common way to move a boat from one elevation to another. However, the flash lock [3,5,7,17,19] is a reasonable starting point, if not from an historical perspective, at least in terms of crudeness or general lack of sophistication. The flash lock, often and perhaps preferably referred to as a stanch, was a device used in relatively small rivers to make them navigable over a greater length or to overcome shallows. They consisted of a wooden frame with a fixed sill and moveable upper beam providing a base for a number of nearly vertical paddles or planks held in place by the water itself. A supported wooden footbridge was used as access for a lockkeeper or boatmen. On occasion the flash lock became fairly sophisticated.

In place, the stanch provided as much as 1.5 m (5 ft) of backwater thereby improving the upstream navigation. Rapid removal would create a "flash" carrying boats through the lock, and across any downstream shallows. [6,19] Obviously, movement against the current was considerably more difficult, and usually required winching. Although the stanch backed up water that would provide power for riparian mill owners, opening the stanch reduced both head and discharge, bringing the boatmen in conflict with the interests of the miller.

The flash lock was a reasonably common device, primarily in the nineteenth century. There were used in the Fens and on the Thames, Severn and Warwickshire Avon, as well as several French rivers. They were removed from the River Severn in 1842 as part of a river improvement program, but lasted on the Thames until 1937 and on the Avon until 1961. A detailed and fascinating history of navigation on the Thames was published by Skempton [16]. Although there is some similarity between the flash lock and the pound lock, the primary purpose of the flash lock was to provide adequate depth rather than an elevation change.

POUND LOCK

The pound lock, frequently referred to hereafter as simply a lock, has always been and remains the primarily means of moving a boat from one elevation or level to another. The pound lock is generally a rectangular chamber created by gates on either end, and some means of letting water in and out of the chamber so that the water level can be varied between the two canal levels.

The Grand Canal of China discussed previously was regularly expanded and repaired. It is reported that during a rebuilding in 984 AD, Chhaio Wei-yo replaced a double slipway over which barges were hauled, by two gates located 46 m (150 ft) apart. If this is true, it is likely to be the first, albeit primitive, pound lock.

However, a primitive double-gated chamber dating from 1180 at Dammes, Belgium on a canal connecting Bruges with the North Sea is frequently acknowledged as the first pound lock. Further, a lock built at Vreeswijk, The Netherlands is considered to be the prototype of the modern pound lock. [7] Along with a multitude of military and civil works, canal systems also fell within the scope of Leonardo da Vinci. [7,13,17] This included plans for a large canal from Florence to the sea that bypassed an unnavigable stretch of the Arno River. He is also credited with the invention of mitre-gates for lock chambers in the late 15th century, previous lock chambers generally having vertical gates.

There are two significant early canals in France. [17] The Briare Canal (1604-1642) that connected the Loire and Seine River systems was the first canal in western world to cross a watershed. This effort required some 41 locks in a length of 55 km (34 mi). The second canal was the Canal du Midi (Chief engineer, Pierre Paul de Riquet). This famous canal contained 119 locks in 238 km (148 mi). One 152 m (500 ft) tunnel was also required. The tunnel was the first to extensively use gunpowder for blasting. It was opened 1661.

LIFTS AND INCLINES

A large number of locks inevitably meant delays and water supply problems, especially at the summit level. This became apparent during the early days of the British canals as indicated by the early introduction of incline planes and water saving devices. Many proposals were put forward with solutions to overcome changes in elevation. [8] Mechanical cranes were tested. On the more speculative side, Robert Fulton [9] wrote a treatise on canal navigation. Therein he included machinery to cross mountainous terrain without the use of locks. Many of his ideas were impractical, but may have influenced other more practical developments.

Great Britain

The narrow canals of England and Wales have a very high concentration of locks necessitated by the hilly countryside. This, along with their small size contributed to the decline of commercial usage of the canals. However, it is these same features that make the canals so attractive today. There were a large number of inclines on the British canals of which the first was built in 1778 at Ketley. [8] The Bude Canal, using small barges fitted with wheels, made exclusive use of inclines for elevation changes. Other canals with significant inclines in the West Country included the Torrington and Chard canals. [4,13,19] All told, there were some twenty inclines on the West Country canals. Some remains of the Ridd Incline may still be found.

Elsewhere, incline planes were used on the Shropshire Canal and on the Monkland Canal at Blackhill, Glasgow. The latter incline overcame a vertical rise of 29 m (96 ft) in a horizontal distance of 317 m (1040 ft). [13] Almost all of the inclines have long ago disappeared from the scene. The most famous today are undoubtedly the Foxton Incline and the Hay Incline at Coalbrookdale. [5,14] The Foxton Incline was completed in 1900 to reduce delays at the Foxton Locks. The locks consisted of two staircases, each of five locks, for an overall elevation change of 23 m (75 ft). The Foxton Incline fell into disuse as early as 1910 because of mechanical problems and the high cost of operation, and was scrapped in 1928. It has been suggested that if the Foxton Incline had originally been electrified, it might well have survived. [13]

The remaining incline at Coalbrookdale (originally one of five) with an elevation change of 63 m (207 ft) in 305 m (1000 ft) saved the expense of 27 locks and required only three and one-half minutes per transit. [5,8] This incline was designed for special tub boats, as were the Bude and most of the West Country canals. Other then their complexity, the major problem with steam-powered inclines was that their commercial viability dropped rapidly as the number of boats decreased. Efforts have been made to partially restore the Foxton and Coalbrookdale inclines and one day they might well be back in operation.

Lifts were less common than inclines, and only one remains in England at this time. Earlier, there were eight on the Grand Western Canal alone. Vernon-Harcourt [19] estimated that utilising the lifts on the Grand Western Canal required only one-third of the time necessary to pass through an equivalent set of locks, and reduced the expenditure of water by 92% in the process. The only surviving lift in Great Britain is the well-known Anderton Lift that vertically raises and lowers boats a distance of 15 m (50 ft) between the Trent and Mersey Canal and the River Weaver Navigation in Cheshire. Built in 1875 to be steam powered and hydraulically balanced, it was rebuilt in 1908 to operate with electrical power. Recently, it has had a chequered history and has been inoperative since 1982. However, every effort is apparently underway to put it back in operation.

Most of the above devices, including the Foxton Incline used water-filled containers or caissons to carry the boats. Other inclines such as those in the US and Poland, both to be described below, transported boats on a carriage with wheels. In addition to the advantage of a nearly hydraulically balanced system, boats continually supported in water did not require as much strengthening. [8]

United States

The Erie Canal, constructed during the period 1817-1825, was certainly the most important of the early American canals. Although a major undertaking for such a young nation, it was rather conventional in its design. However, in the fledgling US, two canal systems are particularly deserving of our attention with respect to the thesis of this paper. They are the Allegheny Portage Railroad in Pennsylvania and the Morris Canal in New Jersey.

When the Pennsylvania Main Line Canal from Columbia, Maryland to Pittsburgh, Pennsylvania was begun in 1826, railways were already on the horizon, and the National Road had already been completed from Baltimore through Maryland and south-western Pennsylvania to Wheeling on the Ohio River. However, the Union Canal (which could claim the first two canal tunnels in the US) was well on its way in eastern Pennsylvania, and the citizens of Pennsylvania could see that the new Erie Canal to the north would have a major impact on east-west trade.

The Allegheny Portage Railroad [11,15,20] was one of the most unusual of transportation systems. Ground was broken in 1826 for the Pennsylvania Main Line Canal with almost no thought given as to how to get past Allegheny Mountain, the high point in the cross-state canal. A 6.4 km (4 mile) long tunnel was suggested although the longest existing tunnel in the US was 259 m (850 ft), and little or no consideration had been given to the summit water supply. The resolution of the impending impasse was more of a railway solution than a canal. Canal boats were transported between water links by railway and the elevation changes were overcome by incline planes. The incline planes were really part of the railway portion of the system, but they transported sectional canal boats. Hence, our interest at this point.

The 444 km (276 mile) canal system built to connect Philadelphia and Pittsburgh consisted of two railroads, the Allegheny Portage Railroad and the Philadelphia and Columbia Railroad, and three canal divisions. In addition to the inclines, there were 66 pound locks roughly 27.4 m (90 ft) long, and 4.6 m (15 ft) to 5.2 m (17 ft) wide. The typical canal cross-section had a 12.2 m (40 ft) top width, a 8.5 m (28 ft) bottom width, and a 1.2 m (4 ft) depth. Originally, passengers (an important part of the system) and freight were transferred between the railway and the canal systems, but after 1834 three or four-section boats were developed that could be disassembled and carried on the rail portion. Each of the boat sections was watertight in its own right. [15]

The solution to Allegheny Mountain was the Allegheny Portage Railroad. The railway was built between Hollidaysburg on the east and Johnstown on the west. (It was the 1889 failure of the South Fork Dam, built to supply water for the western division canal, that led to the famous Johnstown Flood and the loss of 3000 lives.) The railway itself had a length of 59 km (36.6) miles, over which a total elevation change of 783 m (2570 ft) was required; most of the elevation change was achieved by some ten incline planes, the greatest of which was 94 m (308 ft). The summit was 427 m (1400 ft) above the Hollidaysburg end. A single-track was completed in 1834, and a second track added in 1835.

The inclines were powered by stationary steam engines. The cars and boats were originally drawn by endless hemp rope, but subsequently they were replaced by safer steel cable developed by John Roebling. Incidentally, Roebling designed a 347 m (1140 ft) long suspension aqueduct to carry the canal across the Allegheny River and into Pittsburgh. (This was a replacement for an earlier wood structure.)

The inclines had slopes ranging from six to ten percent. The lengths of the inclines ranged from 457 m (1500 ft) to 948 m (3100 ft). A speed of roughly 6.4 km/hr (4 mi/hr) was intended. Between the inclines, rail cars were towed by horse on the short levels, and by steam traction on a longer 22.5 km (14 mi) level. In 1857, the entire Pennsylvania Canal was bought by the Pennsylvania Railroad, and the Allegheny Portage Railroad rapidly disappeared. Portions of the canal were abandoned in 1867, and by the turn of the century the canals themselves were virtually derelict. During its heyday, however, it was a successful canal as well as a popular tourist attraction. The rather spectacular journey was described by many including Charles Dickens.

The Morris Canal [10,12] in New Jersey was originally constructed during 1827-1831. Originally built for boats of 25 tons, it was enlarged for 70-ton barges in 1840-1845. The original dimensions were 9.8 m (32 ft) and 6.1 m (20 ft) for the top and bottom widths respectively, and the depth was 1.2 m (4 ft). The canal climbed from the Delaware River at Phillipsburg to a summit level of 278 m (913 ft) for an elevation change of 231 m (758 ft). It then dropped 278 m (913 ft) to the Hudson River. Since the overall length was 158.7 km (98.6 mile), this required an average grade of 3.4 m/km (18 ft/mile).

The necessary two locks per 1.6 km (1 mi) were felt to be prohibitively expensive so inclines were proposed instead. A series of 23 inclines were constructed under the design and supervision of British engineer, James Renwick. For comparison purposes, it is interesting to note that at the end of the 19th century, it was estimated that there was a lock every 2.21 km

(1.37 mi) on the British canals. [13] Incline grades on the Morris Canal varied between 8 and 10%. The longest incline had a lift of 24.4 m (80 ft), whereas the average was 119.5 m (64 ft). The inclines were powered by Scotch-type turbines and could be operated by one man plus the boat crew. After a period of prosperity, fortunes changed, and the Morris Canal was on the decline by the 1870's. Traffic continued to decrease and the canal was finally drained in 1924. Little remains today.

Northern Europe

A larger scale has made commercial canal transportation in Northern Europe more viable than in the early US and British canals. Correspondingly, lifts continue to be important. A few examples are listed below.

An early German lift near Waltrop, opened in 1899. As the largest structure on the Dortmund-Ems Canal, it surely qualified as a technical engineering wonder of its day. This lift closed in 1970, was converted to an historic technical monument in 1979, and ultimately opened as a museum in 1992. It was subsequently recognised an "European Museum of the Year."

The Oder-Havel (originally Hohenzollern) Canal was built between 1907 and 1914 to move goods between the Havel and Elbe and the Oder and Baltic Sea north-east of Berlin. A number of lock chambers were required to achieve the lift of 36 m (118 ft). Lockage time was unacceptable and the rather amazing Niederfinow Lift was constructed from 1927 to 1934 some 55 km north-east of Berlin to bypass the locks. The lift structure has a total length of 94 m (308 ft) and a height of 60 m (197 ft). The lifting trough is 85 m (279 ft) long, 12 m (39 ft) wide, and the water depth is 2.5 m (8 ft). The elevation change of 36 m (118 ft) requires 20 minutes replacing the much longer locking process. The ingenious counterbalanced system requires only four 75-hp motors to overcome primarily friction and inertia.

A third German canal lift opened in 1938, and still in operation, near Magdeburg raises and lowers boats between the River Elbe and the Mittelland Canal. The lifting chamber has a length of 85 m (279 ft), a width of 12 m (39 ft), and a lift of 16 m (52 ft). The chamber is driven by four electrically-powered large drive screws. They are hydraulically assisted by a below-canal 10 m (33 ft) diameter by 36 m (118 ft) long buoyant tank housed in a 70 m (230 ft) long cylinder. [18]

Belgium also has an fascinating canal history. On the original Canal du Centre between La Louvière and Thieu there are a series of four lifts that are now identified as a World Heritage Site. [21] The system of lifts was constructed between 1888 and 1917 to provide a connection between the Charleroi-Bruxelles Canal and the Mons-Condé Canal. The total change in elevation is some 67 m (220 ft), and the capacity is 300 t (330 ton). The site is an outstanding example of a late 19th century industrial landscape, and was so recognised by UNESCO.

Construction to increase the capacity of the Canal du Centre to 1350 t (1485 ton) began in 1963. In addition to a new channel, a single shiplift at Strépy-Thieu with two 112 m (367 ft) by 12 m (39) caissons was designed to lift the boats some 73 m (240 ft), thereby replacing the original four lifts. Work on the lift commenced in 1982, and was scheduled for completion in 1999. [2]

On the branch to Charleroi, Belgium, an incline plane was constructed at Ronquières in 1967 to replace over 30 locks. This incline plane consists of large water tanks transported on rails. The elevation change of 65 m (213 ft) is obtained in a length of 1500 m (4920 ft).

Larger canals may have their best chance of survival as a commercial carrier, while smaller canals are much more likely to owe their continued existence to leisure boating. In fact, the slower development of boating holidays on the French canals as compared with Great Britain is at least due in part to the larger scale of the French canals. On the other hand, an exception defying any logical economical analysis is the Ostróda-Elblag Canal in Poland. [1] The canal system has an overall length of 187 km (116 mi) including lakes and natural waterways. The main stretch of canal is between the cities of Ostróda and Elblag, a distance of 80 km (50 mi), again including lakes and a short length of river.

Like many other small canals the transport of goods has long since ceased to be of importance, survival is due to a local love for the canal, its spectacular scenery and birdlife, and in this case most importantly, a series of five inclines. Here the five inclines cover a total vertical distance of 99.5 m (326 ft). The lifts range from 13.0 m (43 ft) to 25.5 m (84 ft), and the lengths of the inclines range from 350 m (1150 ft) to 550 m (1800 ft).

The entire canal was heavily damaged during World War II, but by August 1945 the canal was back in business transporting timber. The first tourist boat passed through in 1947, and this has now become virtually the sole function of the canal. What remains however, is a canal incline system that looks and operates in many respects almost identically to that of the Morris Canal some 150 years ago. It is extremely rare today to find examples of technology, especially on this magnitude, that continues to operate essentially as did the technology of the 1800's.

FINAL COMMENTS

In the spirit of this conference, it is hoped that these few remarks will help tie together a few aspects of the long and colourful history of canals. As we enter the new millennium, we can rest assured that historic canals are going to continue to survive, in some cases derelict canals will be put back into service, and canals in general are going to prosper. Great Britain has shown the way, let us hope that other countries will do as well.

This brings us full circle as it were, to a brand new technology as far as canals are concerned: The Millennium Wheel, Falkirk Wheel, or what is apparently its final name, the Antonine Wheel. Since the Antonine Wall was the first physical connection between the Forth and Clyde back in 142 AD, this seems a very appropriate name.

REFERENCES

1. Baldowski, Jan, "The Ostróda-Elblag Canal," Tessa, Poland.

2. Brouet, H., "The Shiplift at Strépy-Thieu, Canal du Centre, Belgium," *Proceedings*, Institution of Civil Engineers, vol. 123, August 1997, p. 178.

3. Buchanan, R.A. *Industrial Archaeology in Britain*, Penguin, Harmondsworth, England, 1972.

4. Cadbury, George, and S.P. Dobbs, *Canals and Inland Waterways*, Sir Isaac Pitman and Sons, London, 1929.

5. Cossins, Neil, *The PB Book of Industrial Archaeology*, David & Charles, Newton Abbot, 1975

6. Cubitt, Sir W., *Minutes of Proceedings*, Institution of Civil Engineers, vol. iv, 1845, p. 111.

7. *Encyclopœdia Britannica*, Chicago, 1998.

8. Evans, Francis T., "Roads, Railways, and Canals: Technical Choices in 19th Century Britain," *Technology and Culture*, vol. 22, 1981.

9. Fulton, Robert, "A Thesis on the Improvement of Canal Navigation," Taylor, London, 1796.

10. Gorman, Mark, "History of the Morris," (Unpublished Report), 1979.

11. Hoffmann, Philip J., *A Picture Journey Along the Pennsylvania Main Line Canal 1826-1857*, American Canal and Transportation Center, York, PA, 1993.

12. Lee, James, *The Morris Canal: A Photographic History*, American Canal and Transportation Center, York, PA, 1974.

13. McKnight, Hugh, *The Shell Book of Inland Waterways*, 2nd ed., David & Charles, Newton Abbot, 1981.

14. Nicholson *Ordnance Survey Guide to the Waterways* (multiple volumes), 1985.

15. Shank, William H., *The Amazing Pennsylvania Canals*, American Canal and Transportation Center, York, PA, 1981.

16. Skempton, Sir A.W., "Engineering on the Thames Navigation, 1770-1845," *Transactions*, The Newcomen Society, vol. 55, 1985.

17. Upton, Neil, *An Illustrated History of Civil Engineering*, Crane Russak, New York, 1975.

18. Various Internet sites.

19. Vernon-Harcourt, Leveson Francis, "A Treatise on Rivers and Canals," Clarendon Press, Oxford, 1882.

20. Welch, Sylvester, "Report on the Allegheny Railroad," Report to Samuel Jones, Esq., Superintendent of the Western Division of the Pennsylvania Canal, 1833 (reprinted by American Canal and Transportation Center, York, PA, 1979).

21. "World Heritage Sites," *Report*, UNESCO, World Heritage Committee, 22nd Session, 1998.

The Shannon-Erne Waterway

DR. THOMAS BREE,
ESB International, Dublin, Ireland

INTRODUCTION

Ireland has a wealth of waterways - rivers, lakes and canals - with no commercial traffic to disturb their sense of peace and serenity. You can plot your route to suit your own tastes - along the magnificent Shannon, the beautiful Barrow, the island-studded lakes of the Erne, the tranquil Grand Canal or the changing face of the Royal Canal which is in the process of being restored.

The Shannon-Erne Waterway comprises 63 km of lake, river and stillwater canal and was re-opened in 1994 at a cost of £30m. It allows navigation for the first time in 125 years between Ireland's two great waterways, the Shannon and the Erne. It flows past sixteen automated locks, thirty one stone bridges and six fully serviced mooring areas. The Shannon-Erne Waterway, renamed from the original Ballinamore-Ballyconnell Canal, offers glimpses of a part of the country not often seen by visitors. You can meander through a variety of country landscapes, enter lakes where you are the only boat for kilometres, or visit the charming towns and villages along the waterway.

The three-year project was a major joint undertaking, financed by the Irish and British Governments, EU Structural Funds, the International Fund for Ireland and Electricity Supply Board - they operate water management systems on the Shannon and Erne. ESB International managed the construction and it is maintained and operated as part of the integrated system by Waterways Ireland, a joint North-South Body.

The Millennium Link. Thomas Telford, London, 2000, 152–158

THE ORIGINAL SCHEME

William T. Mulvaney, one of Ireland's greatest engineers of the 19th century, undertook a feasibility study for the original project. He was later to immigrate to Germany where he became a founder of the coal mining industry and is remembered today by a statue in Dusseldorf.

It had always been envisaged that the northern and southern navigation networks would eventually be connected. An independent stillwater canal was recommended to the Commissioners for Public Works as the most efficient way of making the link. However, under pressure from local landowners to reduce flooding in the Woodford River area, the Commissioners decided to kill two birds with one stone by carrying out a combined drainage and navigation scheme. They gave the task to John McMahon, a talented engineer, and work began in 1846. McMahon planned his route in three sections:

- A connection to the Shannon at Leitrim village, made by converting the tiny Leitrim River into a Stillwater canal. This involved considerable excavation to create a stairway of eight locks rising from the Shannon to Lough Scur.
- A channel joining Lough Scur and nearby Lough Marrave to provide the summit reach, a 4.8km stretch of level water which would carry boats across the high ground separating the Shannon from the Erne.
- Canalising the Woodford and Yellow rivers to provide the remaining 45km of the 62.5km navigation, reaching the Upper Erne near Belturbet.

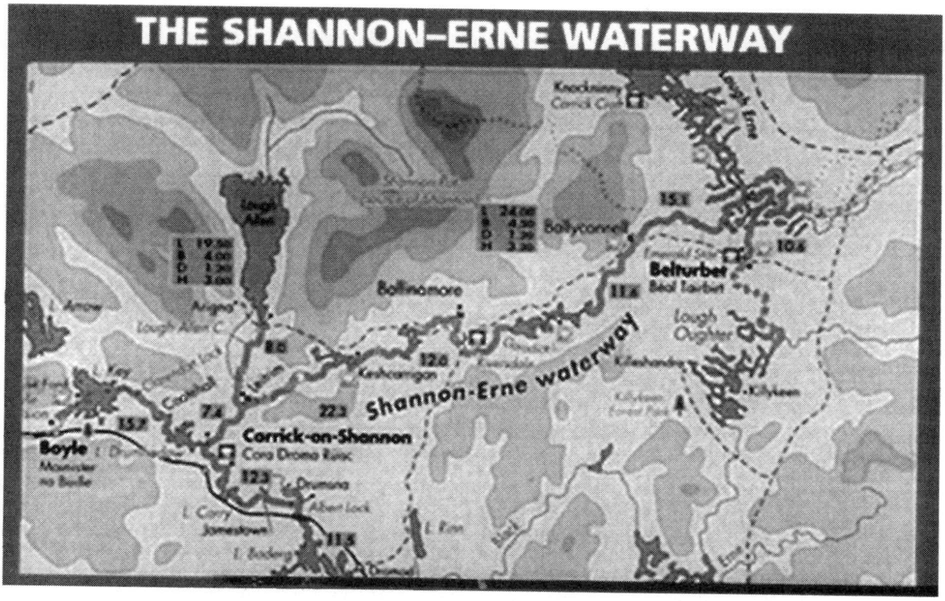

The conflicting interests of drainage and navigation inevitably led to compromises and McMahon's project became increasingly difficult as time wore on. The 4,000 workforce he employed at the outset dwindled away to 300. This unexpected shortage of workman greatly slowed progress and constant floods caused further delays.

On 4th July 1860, the canal was finally completed.

It was officially handed over to twelve trustees, who were soon complaining about banks caving in, leaving blockages at locks and other deflects caused by cost cutting during construction. Meanwhile, the advance of the railways had undermined the value of waterway connection between Belfast and Limerick. Now that the link was at last a reality, it was questionable if there would be significant demand.

In all only eight boats passed through the Ballinamore and Ballyconnell Canal and paid tolls during its working life - fewer than one a year! The waterway, which took four years to plan and seventeen to build, linked the Shannon and Erne for only nine short years before it was abandoned in 1869. On the other hand, McMahon's flood-prevention measures have proved totally successful.

After 120 years of neglect the canal became a sad picture. There were no gates left on any of its 16 locks, the stonework of the lock chambers, quays and bridges had been damaged by the unchecked rush of water, and even experienced canoeists found negotiation of the clogged and weed-grown channel challenging.

RECONSTRUCTION PLAN

The waterway authorities undertook preliminary surveys in the 1980s and in April 1989, the Governments decided to carry out all necessary economic and technical assessments and to put together a funding portfolio.

The Government Agencies were represented by a Joint North/South Steering Committee, with four government departments from each side. Many additional and local government bodies were consulted on a range of issues and the complete detail design was issued for comment to six agencies. Further consultations were held with farming organisations, waterways associations and other voluntary bodies.

The economic benefits of the scheme were identified through consultations with foreign cruise hire operators and experiences in cruising holidays at home and abroad. The gross benefit of the project over a 25-year period was estimated at £45m excluding secondary benefits from the use of the canal for angling, canoeing, walking along the towpath and other tourism activities in region.

Channel dimensions were proposed on the basis of a questionnaire survey of all cruise-hire operators and members of the Inland Waterways Association. Conditions on the Shannon and Erne systems were reviewed and site tests were made to assess cruiser-manoeuvring capacity in a narrow canal.

KEY FEATURES

- Constructing embankments, with weirs and by-pass channels, between the Shannon and the summit at Lough Scur. Deepening and widening sections of the Woodford River channel between Lough Scur and the Erne. Protecting banks with rock –armour where necessary and with timber piles in soft ground.

- Carrying out the work from one bank only, where possible and using floating plant to say clear of the land altogether in environmentally sensitive areas, so that natural habitats are preserved.

- Major dredging of Ballymagauran Lake and Upper Lough Erne, plus minor dredging of the access to five other lakes, with transportation of spoil to designated land-based areas.

- Replacing ten bridges, and underpinning and making major repairs to a further 24.

- Reconstructing eight lock chambers using existing masonry and repairing the remaining eight. Providing all sixteen locks with new automated gates and sluices. Rebuilding weirs and fish passes at all locks on the Woodford River section.

- Installing pumps at each of the eight locks leading from the Shannon and at one lock on the Woodford River, to pump water up to summit level. Investigation and lining of impermeable limestone in a rock cut.

- Providing six public mooring areas with berthing facilities, sewage pump-out facilities, water supply, toilet block with shower and laundry room, refuse disposal.

- Landscaping and planting along the worked riverbanks and mooning areas.

- Electrofishing and removal of fish to safe havens during the works and restocking on completion.

- Provision of services to adjoining lands, such as water supply, drainage, access, etc.

- A summary of the major quantities is shown in table 1.

Table 1. Major Engineering Quantities

Soft Excavation	800,000 cu. m.
Rock Excavation	150,000 cu. m.
Concrete	25,000 cu. m.
Structural Steelework	220 tonnes
Access Roads	approx 5.5km

NAVIGATION STANDARDS

It is well known that, at certain speeds, boats travelling in restricted waters experience a rapid increase in resistance. Dramatic changes occur at the water surface and take the form of a rapid increase in the drawdown alongside the boat, coupled with a sudden growth in the height of the waves aft of the boat.

The onset of limiting critical conditions was examined through calculations, which were calibrated by a prototype test on site. The comfortable speed, from the point of view of boat control and manoeuvring, was taken at 90% of the limiting speed, giving the "operational condition".

A minimum depth of water of 1.55m was chosen, allowing 1.2m for the boat draft and 0.35m for squat and weed growth. The limiting speed is 3.4 knots for large cruisers in a channel width of 13m. This also allows two cruisers of beam 3.8m each to pass comfortably. On many sections of the canal and in the lakes the channel dimensions will be larger.

At bridges the minimum airspace over water level during non-flood periods is 3.65m over a 3.0m width. Happily, this caters for 96% of all cruise hire craft without the necessity to reconstruct existing masonry arch bridges. It provides navigation in summer floods up to those with a return period of five years. In some cases the water level and channel bed was lowered rather than interfere with the fine old masonry structures. Where existing concrete bridges had to be raised, headroom varies from 4.0 to 5.5m. The relatively high figure of 5.5m applies at a bridge on Lock No. 1 at the northern end, where water levels can be high for long periods due to the backwater effect from Upper Lough Erne.

ENVIRONMENTAL IMPACT

The Environmental Impact Assessment Report listed five significant issues:

- Socio-economic factors
- Drainage
- Water Quantity and Quality
- Ecology

- Landscape

The Shannon - Erne Waterway is of international importance for wintering wildfowl, and holds an internationally important number of Iceland breeding Whooper Swans, and lesser numbers of both dabbling and diving duck. Extensive surveys were made of wintering wildfowl and of breeding waders.

An ecologist was retained during the construction period to monitor the works.

At significant areas of ecological interest, located mainly close to lakes, the dredged silt was removed from the waterway banks by barge. Outside these areas of special interest, work was carried out from one bank only. In areas of stiff clay and silt the unworked bank was untouched, preserving both aquatic and marginal flora. In very soft areas it was necessary to have gently angled side slopes throughout.

STRUCTURES

There are sixteen locks on the Shannon - Erne Waterway. A stairway of eight locks, located within five kilometres of each other, raises the canal from the River Shannon, near Leitrim, to the summit at Lough Scur. The other eight locks are spread along the canalised Woodford River, as it flows down the summit to Upper Lough Erne.

The locks on the Stillwater Section were in reasonable condition at the start of the project. It was concluded that these locks could indeed be repaired and that the lock chambers were of a size suitable for use by modern cabin cruisers, i.e. 25m long x 5m wide x1.7m deep.

A decision was taken at the start of the project that none of the masonry arch bridges should be replaced. However, work was required on all these bridges to strengthen them and provide room for navigation. Towpaths were generally removed from under these bridges so as to widen the narrow navigation channel through them.

Many of these bridges were underpinned, as canal bed levels were lowered and a concrete paving slab was provided in the river bed to prevent scour under abutments, which were not founded on rock. Other repair works on these bridges included the removal of vegetation, cleaning down of masonry, replacement of missing masonry blocks, repointing the masonry and grouting of the fill material behind the abutments.

RURAL REGENERATION

The project has been an outstanding success in terms of economics, environment and heritage. The return on investment has been proven, based on the revenue generated from boat hire, private boats, day boats, anglers, cycling, walking, special interest groups and other complementary developments. Within three years, there was a 34% increase in guest bedrooms in the region and this number continues to rise.

Table 2. Number of Boats on the Waterway

1994	3300
1995	3600
1997	3900
1998	3700
1999	3500

The number of boats on the waterway is shown in Table 2. The high level of interest is related to the strong marketing effort - initially a budget of £3m was allocated. The recent numbers have declined slightly due to the decline in the German economy; one third of the boat-hire tourists are from Germany. However, they are still higher than the first year figure.

The success of the project has rekindled an interest in waterways and rural regeneration for many other parts of the network in Ireland. Work is underway on the Royal Canal and on four minor schemes, while a major programme has been developed for the part of the network in Northern Ireland.

Towards a new golden age of waterways?

DAVID EDWARDS MAY, EURO MAPPING

The restoration of old canals and river navigations has become an issue of regional development in many parts of Europe. What used to be a relatively marginal activity of a small group of enthusiasts devoted to conservation of what was left of our waterway heritage has blossomed into a specific sector of economic activity in the areas concerned, showing a range of benefits sufficient to justify the substantial investments involved.

At the macro-economic level, Europe still needs new waterways to complete its high-capacity network for modern inland shipping, and indeed every 5 or 10 years a vital new link is opened (the Danube-Black Sea Canal in 1985, the Main-Danube Canal in 1992, the Saar in 1995). At the local level, however, 'heritage' waterways are proving that they too have a major contribution to make to economic development.

Hence the Millennium Link project which is the subject of this conference, but it is important to note that this is far from being an isolated example of large-scale engineering being applied to the revival of historic waterways, for the unique combination of social, economic and environmental benefits they offer.

This paper looks at how engineering, economic and cultural aspects are interconnected in the examination and implementation of waterway projects on the Continent. The emphasis is on the smaller heritage waterways, but such is the commonality of purpose and engineering challenges between these waterways and their modern counterparts that we prefer to consider the European waterways as a whole. Between the Forth & Clyde and the Mittelland Canal in Germany, only the level of service differs. Essential principles are the same: level pounds, structures to overcome the differences in level between them, and the most rational and sustainable mode of transport for people and goods.

RESTORATION - TWO PHASES OF POST-WAR DEVELOPMENT

The movement to put disused waterways back on the map has seen two phases since the Inland Waterways Association started its campaigns in 1946, and this conference perhaps marks the beginning of a third. The pioneering phase made its breakthrough in 1964 when the Southern Stratford Canal was re-opened by the Queen Mother. It continued through to the 1980s, the supporters of long-term projects such as the Kennet and Avon Canal showing remarkable determination and patience in the absence of major public funding.

The current phase corresponds to widespread adoption of waterway restoration projects by local and regional councils working partnership with the waterway authorities, with much less effort required on the part of the volunteer movement. The change took place in the UK when British Waterways committed itself to the complete restoration of the Kennett & Avon in the mid-1980s, but at about the same time a new public awareness of the potential value of disused waterways was appearing in France.

Seeds for the first major waterway restoration project were sown in the early 1970s, when riparian communes and départements of the river Lot in South-West France formed the Association pour l'Aménagement de la Vallée du Lot, to tackle problems of depollution following abandonment of the coal workings around the upstream limit of navigation near Decazeville. At 256 km long with about 70 locks, the Lot is one of the longest canalised rivers in Europe. It had been abandoned as a commercial navigation nearly 50 years earlier, leaving a void which the national electricity undertaking Electricité de France was quick to fill, by building imposing hydropower schemes at five sites along the river: Cajarc, Luzech, Fumel, Villeneuve and Castelmoron (see illustration).

These structures, with a fall of between 6 m and 13 m, were considered insurmountable obstacles, but the association's president Christian Bernad felt that the river had great potential for tourism even if only isolated sections were restored. He was encouraged by the growth in the hire boat industry on French waterways, particularly rapid in the years 1980–85, and by the pioneering operation of a cruiser on the long reach impounded by Luzech dam, an initiative promoted by Cahors-based tourism development officer Jean De Chalain. Further evidence was provided by a feasibility study commissioned by Bernad's association, the Cahors Chamber of Commerce and the Conseil Général in 1986.

The key finding, relevant also for the Lowland canals, was that isolation from the national waterway network was no obstacle to development of a successful tourist product, although the hire boat activity would clearly be preponderant. What hire firms in France and the UK indicated in response to the 1986 survey was that the envisaged length of 37 km was too short, and that 46 km would be a minimum for a credible one-week hire boat holiday. This meant building a completely new lock at Galessie, because of a micro-hydropower plant installed in the former lock chamber, owned by the treasurer of the Chamber of Commerce!

Enthusiasm for the project snowballed after the project's protagonists had won the political support of Maurice Faure, a French government minister and president of the Conseil Général. Renowned for his skepticism, he was won over by a carefully engineered slide presentation which showed what the river would look like if boats returned! It was then found possible to extend the first phase of restoration to a total of 64 km, giving access to the hilltop village of Saint-Cirq-Lapopie (64 km).

Thus in 1990, the first substantial waterway restoration project to be carried out in France was completed at a cost approaching £ 4 million, with 13 locks restored and one built from scratch.

The findings of the feasibility studies for this project were confirmed, with three major hire firms setting up bases, as well as a smaller local firm and three trip boats: all this on a navigation unconnected with the other navigable waterways in south-west France!

The département of Lot et Garonne, which holds the key to interconnection with the Canal du Midi route across southern France, is now completing restoration of the entire 83 km of the river Lot in its territory. This required construction of a completely new 10m deep lock at Castelmoron (beside the EDF dam), completed in 1996 at a cost of £3 million. Works are now in progress on the 13m deep lock at the Villeneuve dam.

Complete restoration of the Lot throughout the 256 km was estimated in 1992 at £40 million, but this is already known to be inadequate in view of major difficulties identified on the Luzech loop. There is also a lack of consensus among the départements concerned on a coherent programme of works. However, with virtually half the length of the river soon to navigable again, it is not unreasonable to forecast that these difficulties will eventually be overcome.

At a conference held in Villeneuve on 24–25 May 2000, Lot et Garonne celebrated restoration (with the neighbouring départements of the Lot and Gers) of more than 200 km of river navigations abandoned between 1926 and 1955 : the Lot, the Baïse and the short length of the Garonne which links the two.

The Finow Canal project in Germany is less ambitious, but is unique in being the first canal restoration project to be promoted in Germany. It also represents a different approach to waterway restoration. Where investment in the river Lot was almost exclusively justified by projected boating use, the Finow Canal Association has secured Federal and Land financing for restoration of this historic canal near Berlin essentially to retain the linear asset and its structures as industrial archeology.

The Finow Canal provided Berlin with its first navigable link between the Havel and the Oder. Built in 1605 to 1620, it was destroyed during the Thirty Years War (1618–1648). It was rebuilt between 1743 and 1746, and comprised 15 locks over a total length of 41 km. By the mid-19th century, canal traffic was so heavy that even 24-hour-a-day lock operation was insufficient to handle all movements, and a second parallel lock chamber had to be provided at each fall. All the locks had a layout on plan that is unique to the waterways between the Elbe and Oder. They were built to accommodate two Finow barges (length 40.20 m, beam 4.60 m, draught 1.60) side by side. The upstream and downstream lock heads were staggered so that the first barge to enter the lock would always be the first one out.

The Finow Canal remained almost unchanged until the beginning of World War I. It was one of the most heavily used waterways in the world, with annual traffic amounting to 3 million tonnes. Even the Rhine was carrying only slightly more traffic.

Between 1906 and 1914 a new Havel-Oder Canal was built running parallel to the Finow, designed for 700-tonne barges. After June 1914 the majority of between the Oder-Berlin traffic began to use the new canal, but a 32 km section of the old canal comprising 12 locks was left intact. However, with traffic now marginal, the second lock chambers were gradually infilled or in one instance rebuilt as a bypass weir for surplus flow. The canal still carried minimal freight traffic until 1973. From that date, however, it began to be used exclusively by boat traffic, and even that was sporadic. All navigation ceased in 1992, following a major failure at one of the locks. Since then, other locks and dams have also fallen into disrepair. In addition to these structural problems, the canal started to silt up.

Foundation of the Association for the Preservation of the Historic Finow Canal in June 1995 marked the beginning of an intensive campaign for restoration of the canal, and the objective has since then been partially fulfilled. On the occasion of celebration of the 250th anniversary of the canal on June 16, 1996, the Brandenburg Government committed itself to maintenance of the waterway, especially as a vehicle for developing tourism in the region.

However, since the waterway belongs to the Federal Government, it remained to obtain ministerial approval from Bonn, to secure the finance required to implement the works. This

On many navigable rivers in France the decline in commercial navigation after World War I opened up valuable potential for tapping the potential hydropower from even modest heads Micro hydropower plants have squatted countless locks, the concessionaries getting particularly favourable deal both from the State and from the national electricity industr Electricité de France. With renewable energy high on the list of the nation's developmen priorities in the post-war period, EDF was not going to let small private businesses exploit a the possibilities. Accordingly, it was given concessions to build dams of modest head, in th range of 6 to 40 m, flooding all the navigation works for a certain distance upstream.

The head at Castelmoron, on the river Lot, is 10m, and the dam was completed in 1955. Restor ation works were proceeding on the river further upstream when the author visited this site

Germany is in the forefront of development of new and upgraded waterways to accommodat high-capacity inland shipping between East and West. At the heart of Projekt 17 is the Elb crossing at Magdeburg, illustrated here, where demolition of the 1930s honeycomb reinforce concrete structure was required to make way for an entirely new aqueduct built to a new an simpler design.

The new 73m high Strépy Lift on the high-capacity Canal du Centre in Belgium towers above the Haine valley, with hydraulic lift No 4 at Thieu in the foreground (head 16 m). This emblematic site of European waterways features engineering works, extant or derilict, spanning two centuries, promoted through a "Regional Canals Park". The message here is that today's engineering achievement becomes tomorrow's heritage.

n 1991, to see whether a by-pass could be envisaged. Apart from a pylon to be removed, and a narrow strip of private land to be expropriated from an unaccommodating baroness (foro the downstream lock approach), there seemed to be no obstacle. Six years later, EDF was proud to be opening this 32 by 5,50 m lock, heralding the eventual restoration of navigation over the entire 260 km length of the navigation.

approval came in June 1997, covering 75% of the projected cost of DM 42 million. The remainder is financed by the Brandenburg Government.

Removal and disposal of contaminated silt (lead and quicksilver are known to be present in significant quantities) alone accounts for about a quarter of the total cost.

The works required to restore the abandoned lock at Stecher, and to secure the other threatened structures, have already begun. The canal structures that are protected as historic monuments will be restored to their original condition. The others will be maintained or adapted according to their functions. This could mean rebuilding some of the lock chambers to smaller dimensions.

The restored Finow Canal has a vast potential market, with tens of thousands of private boats moored in and around Berlin.

NEW LINKS TO COMPLETE THE NETWORK?

The new phase mentioned above, heralded by the Millennium Link, has involved a leap of both imagination and ambition, so that it has become almost reasonable to promote not only the restoration of old routes but the construction of new links to complete the network!

Some such projects have been promoted since the early days of the waterway conservation movement, and have involved bold designs. The earliest was the Upper Avon Link conceived by David Hutchings, engineer of the Stratford and the River Avon restoration schemes. By canalising a few miles of river above the current head of navigation, past Warwick castle, access would be gained to the main line of the Grand Union Canal. The 15m difference in level could have been overcome by a single deep lock with water-saving basins. However, the environmental sensitivities surrounding free-flowing rivers like the Avon are so great that the project may never get off the ground.

Others which seemed more impracticable are now going ahead, such as the Ribble Link and of course the Falkirk junction.

Since completion of the Shannon–Erne waterway in 1994 (involving more new construction than actual restoration), Ireland is discovering the benefits of its integrated waterway network, which is attracting a growing number of private boats as well as hire fleets. With the Royal Canal to complete and the massive Ulster Canal project, one would have thought it premature to be envisaging new links, but that is exactly what is now happening. The Brady Shipman Martin report of 1992 surprised me by its mention of a completely new cut from the Royal Canal west of Abbeyshrule to Lough Rea, opening up an attractive circular route. Now one hears talk of a link from the restored Boyle Navigation up to the Royal Canal, with discussion of the respective merits of boat lifts and locks.

Whether new links will ever be built for tourism and recreational traffic in Continental Europe cannot be predicted, but should not be precluded. However, there is such vast potential for restoration of abandoned lengths and for bypassing obstacles such as hydropower dams, that efforts will probably continue for many years in this direction, postponing the introduction of the hypothetical third phase involving new construction projects. This is sensible when one considers the France has more than 3000 km of abandoned navigations!

Among the restoration projects currently under examination are the Canal de Nantes à Brest in Brittany, the Canal du Rhône au Rhine south of Strasbourg down to the Vauban fortified town of Neuf-Brisach, the Tarn navigation from Montauban to Moissac (opening up a circular route with

'rench engineer Jean Aubert developed in the 1960s the unique concept of the water slope, 'hich was tried first on the Canal latéral à la Garonne at Montech, opened in 1974, then at 'onserannes on the Canal du Midi (illustrated here), opening in 1986, after three years of 'rocedures following an early failure. Quite apart from that unfortunate incident, the 'onserannes water slope poses a number of problems that are symptomatic of the hesitations urrounding the change in vocation of France's smaller waterways (250-tonne capacity) from 'eight to tourism and leisure. It was never used by the barges for which it was designed, and 'roved inappropriate and far too expensive in operation to make it a feasible option to the istoric lock staircase for boat traffic. The saving on water is a plus with the water slope, but not) critical on a canal like the Midi which in any event has to convey a substantial volume of 'ater down to the lowest pounds for irrigation.

'ther reservations could perhaps be made about the water slope on aesthetic grounds, and it ust be admitted that the tarmac-flanked concrete flume cuts somewhat brutally through the ind beside the locks. Some operational buildings dating from the mid-19th century indeed had be demolished. A bold statement, no doubt, but the idiom is more that of the modern motorway ian anything to do with environmentally-sensitive hydraulic engineering.

the Canal latéral à la Garonne), the Canal du Berry ‹ a long-term project in view of the extensiv infilling that took place in the 1960s and 1970s and the Upper Rhône.

All these projects involve new challenges for engineers, especially where hydropower dams hav to be bypassed. The Nantes-Brest connection involves restoration of 52,3 km of canal, originall with 72 locks, but the reservoir impounded by the Guerlédan dam has flooded about 10 of thes (which will incidentally provide some welcome relief for boaters!). The difficulty is bypassin the dam itself. A feasibility study commissioned by Brittany Regional Council in 1997 came u with two preferred options: the transverse inclined plane on the Arzviller model (illustrated) an the marine railway.

On the Upper Rhône, the hydropower schemes completed by CNR as recently as 1987 hav heads of up to 18 metres. Here, too, technical options were studied by CNR and the best solutio was found to be the construction of two separate locks with an intermediate basin.

HERITAGE-FRIENDLY SOLUTIONS?

Part of the new trend in waterway restoration projects is the much greater consideration no given to the heritage value of the waterways and their original structures.

Listing of the Canal du Midi (built 1666–1681) and the Canal du Centre in Belgium as UNESC World Heritage Sites, recognition of the Göta Canal as Sweden's historic monument of th millennium, and the work of the EU-funded 'Living Waterways' partnership, among othe initiatives, have all served to underline the cultural value of the waterways and the way the design evolved over the centuries, starting with the Dutch-inspired projects in the 11th–12t centuries, and marked by four periods of expansion led respectively by Italy (15th century France (17th), Britain (late 18th/early 19th) and Germany (late 19th).

Clearly, the new understanding of these values, and the protection measures now implemented t preserve the environmental and built heritage, would now preclude the treatment inflicted upo the Fonserannes staircase on the Canal du Midi at Béziers (see illustration).

But apart from protecting the old, the philosophy surrounding projects discussed in Europe no embraces, as at Falkirk, a sense of pride in designing new structures, to enhance the heritage.

The new phase mentioned above, heralded by the Millennium Link, has involved a leap of th imagination and ambition, so that it has become almost reasonable to promote not only th restoration of old routes but the construction of new links to complete the network!